バイオ実験 英語でトライ！

留学する人も留学生を迎える人も必見！

バイオ研究者のための基本英会話

村澤　聡（先端医療センター 再生医療研究部）／著
Geoff T. Rupp（Language Resources Ltd.）／英文監修

羊土社

羊土社のメールマガジン
「羊土社ニュース」は最新情報をいち早くお手元へお届けします！

主な内容
- 羊土社書籍・フェア・学会出展の最新情報
- 羊土社のプレゼント・キャンペーン情報
- 毎回趣向の違う「今週の目玉」を掲載

バイオサイエンスの新着情報も充実！
- 人材募集・シンポジウムの新着情報！
- バイオ関連企業・団体のキャンペーンや製品、サービス情報！

いますぐ、ご登録を！（登録・配信は無料） ➡ 羊土社ホームページ　http://www.yodosha.co.jp/

はじめに

　2000年5月，私はボストン郊外にあるタフツ大学医学部に所属するセントエリザベス メディカルセンターへ留学しました．それまで，年1回程度の国際学会のため米国を訪れ，幸い何度か英語で発表する機会がありましたが，当然のごとく？ 一般の日本の英語教育を受けてきた一人だったので，とても満足のいく結果ではありませんでした．長期滞在の留学となると，そこそこ英語も話せないといけないと思い，渡米の1年前から仕事の合間に英会話教室へ通って準備を進めていました．しかし，留学の半年前にラボへ下見に行ったときボスと話す機会があったのですが，このときほとんど会話ができず，このままでは大変なことになると感じてからはさらに積極的に英会話教室へ通うようになりました．そのような準備にもかかわらず，いざラボ登場初日，入室して最初に話したのは日本語でした．事前にわかっていたことですが，私が留学したラボは，日本人の割合が最も多かったのです．

　しかし，研究をはじめるとテクニシャンと話しをする必要も出てきます．試薬の注文の仕方や，機器の使い方を聞かないといけません．日常会話は駄目でもラボ英語は何とかなるかと思っていたのですが，そううまくはいきませんでした．当然伝わると思って尋ねた機器の名前が伝わらなかったり，昔習った文法に忠実な英語が伝わらなかったりしたのです．和製英語を正式名と思っていたり，試薬の会社の名前を日本で呼んでいた名称のまま話して伝わらなかったり，本当にどうしてこんなに伝わらないのか，日本での準備はいったい何だったのかと思う日々でした．しかし，時間が経つと不思議なもので会話のスピードにも少し慣れ，何となく感じがつかめてくるようになりました．そのとき思ったのは，nativeの人が使っている言い回しや，正しい機器の名前や，試薬名を最初からもう少し知っていたら結構伝わるんだなということでした．

　私自身，決して留学中にこれらをマスターしたわけではありませんし，別の考え方をすると留学中のこのような苦労も後になってみればよい思い出だったということになるかもしれません．しかし，留学した当初はやはり日々不安でしたし，少しでもこれらの苦労が緩和されていれば，また違った留学生活を送れたかもしれません．特に，留学が短期の場合は，できるだけ早く

生活をセットアップして，自分の本業を軌道にのせないといけません．そのために，この本がこれから留学しようとする方々にとって少しでも手助けになればと思います．

　本書はできるだけ多くのイラストを加え，対話形式にしました．また，1つの表現にできるだけ複数の言い回しを加えたのも特徴です．つまり，一般に日本人が好んで使用する英語から，nativeの英語まで，いくつかニュアンスの違うものを載せてみました．

　Section Iでは，実験室でよく用いる機器類についていくつかの研究場面を想定し，イラストを多用して構成しました．各機器の部分の名称についてもできるだけ英語表現を付しました．Section IIでは，実験の流れを重視しました．個々の実験の詳細な手技については成書に譲りますが，実験の準備，DNA実験，タンパク質実験，そして細胞培養についての概略が，初心者でも理解しやすいように工夫されています．Section IIIでは，私自身が研究を始めて苦労した点を取り上げましたが，4コマ漫画の形式にして，気楽に読んでもらえるようにしました．海外での研究生活ではうまくいかないことの方が多いですが，その場でDiscussionをして，次に役立てることが大切です．ここに挙げた場面を参考にして，活用してもらえれば幸いです．本文のnative speakerによる校正は，医学英語の翻訳を専門にされているラングエジ・リソーシズのジェフ・ラップさんにお願いしました．研究機器や研究内容について詳細に情報を集めていただいたうえでの校正ですので，必ずや海外で即戦力になる英語であると信じています．本書が若手研究者，大学院生，学生の方々に参考にしてもらえるような書物であることを願っています．諸先輩，先生方からはコメントをいただければ幸いです．

　最後に，本書の発行にご尽力いただいた石田さんをはじめ羊土社の皆様，また留学中にお世話になった方々，日本へ帰ってから校正を手伝ってくれた岡本さん，本当にありがとうございました．また，楽しかったけれど辛いこともあった留学生活中，いつも家で笑って悩みを緩和してくれていた家族に感謝します．

2003年6月

村澤　聡

本書の構成

Section I イラストでわかる実験室での英語表現
>> まずは英単語をチェック！

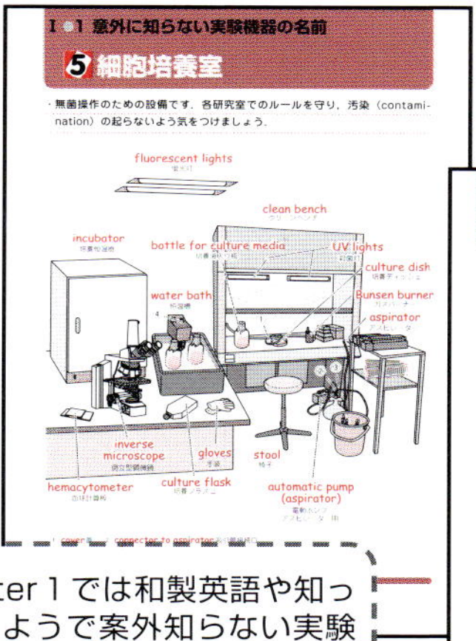

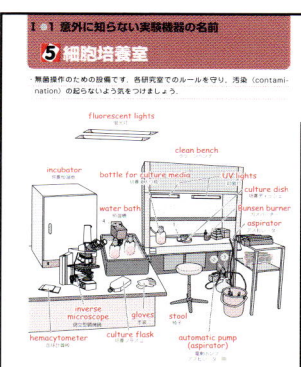

留学先で研究室内を案内してもらっているかのように，イラストを見ながら英単語が確認できます．

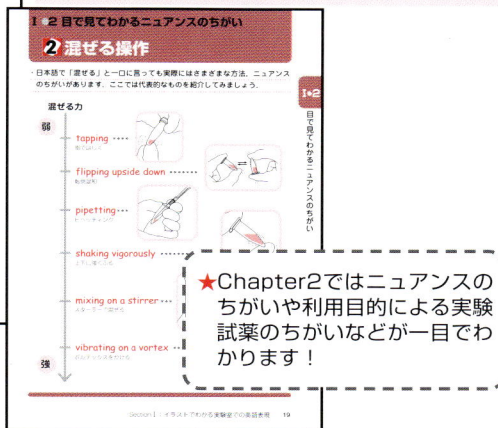

★Chapter 1では和製英語や知っているようで案外知らない実験機器の名前を紹介しています．

★Chapter 2ではニュアンスのちがいや利用目的による実験試薬のちがいなどが一目でわかります！

Section II Let's speak! 〜実験中の英会話
>> 基本的な英会話を覚える！

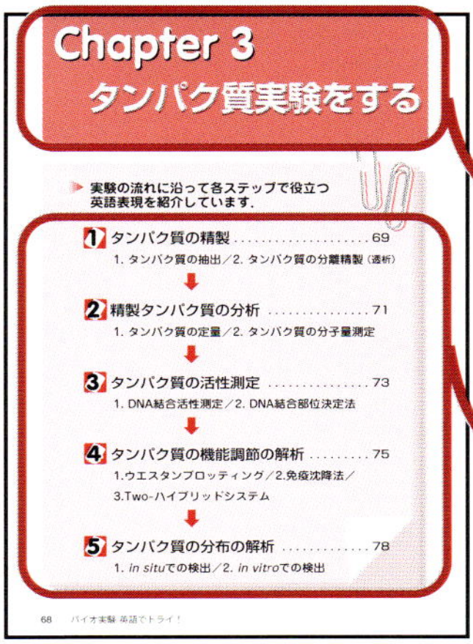

実験前や実験中に，誤解なくスムーズに通じる基本的な英会話を紹介しています！

★DNA実験やタンパク質実験など，大まかな実験の種類ごとにまとめてあります．

★まずは自分の研究に関係する項目から読みはじめてもOK！

★具体的な実験手順に沿って，必要となってくる英会話が習得できます．

次ページにつづく >>>

>>> Section Ⅱ のつづき

★実験操作に関連した機材や試薬などの英単語はここでチェック！

★重要フレーズには，基本的な表現の他に応用表現も紹介しています．いろいろな表現を身につけ，会話に広がりを持たせましょう！

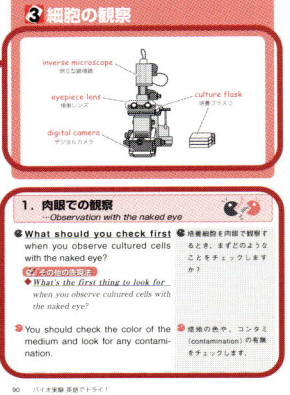

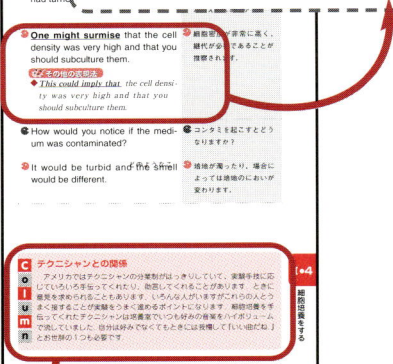

★実験の流れに沿って
🍫（質問者）と 🍥（回答者）による対話型式で会話が進められています．
★実験をシミュレーションしやすく，英会話もどんどん覚えられます！

★アメリカでの体験談が満載！異文化の地での研究生活が垣間みられます．

Section Ⅲ　4コマ漫画でシミュレーション　〜実験結果を報告しよう！
>> ディスカッションにトライする！

4コマ漫画型式なので，臨場感を味わいながら気楽に読み進めることができます！基本的な表現やニュアンスがイッキにつかめます．

★同僚とのやり取りや，機材の予約をする場合に参考になります！

★実験結果の相談やディスカッションをシミュレーションしています．これを参考に自信を持って積極的にボスに話しかけましょう！

バイオ実験 英語でトライ！
バイオ研究者のための基本英会話

Section I ：イラストでわかる実験室での英語表現

Chapter 1　意外に知らない実験機器の名前 ──── 12
- （1）実験台の周辺 ………………………………………………… 12
- （2）計量台の周辺 ………………………………………………… 13
- （3）DNA実験の実験台周辺 ……………………………………… 14
- （4）暗室 …………………………………………………………… 15
- （5）細胞培養室 …………………………………………………… 16
- （6）クリーンベンチの周辺 ……………………………………… 17

Chapter 2　目で見てわかるニュアンスのちがい ──── 18
- （1）水の種類 ……………………………………………………… 18
- （2）混ぜる操作 …………………………………………………… 19
- （3）泳動ゲルの種類と分子量 …………………………………… 20
- （4）泳動ゲルおよびメンブレンの染色法 ……………………… 21
- （5）実験材料の保存 ……………………………………………… 22

Section II ：Let's speak! ～実験中の英会話

Chapter 1　実験の準備 ──── 24
- （1）実験をはじめる前に ………………………………………… 25
  - 1. 実験室での服装 ……………………………………………… 25
  - 2. 実験計画 ……………………………………………………… 26
- （2）試薬／器具の準備 …………………………………………… 27
  - 1. 試薬の準備 …………………………………………………… 27
  - 2. 器具の準備 …………………………………………………… 28
- （3）試薬の取り扱い ……………………………………………… 30
  - 1. 試薬の調整 …………………………………………………… 30
  - 2. 試薬の保存 …………………………………………………… 31

（4）実験を終えて……………………………………… 32
    1. 廃液処理……………………………………………… 32
    2. 実験機具の扱い ……………………………………… 33

Chapter2　DNA実験をする ── 36

（1）DNAを抽出する…………………………………… 38
    1. エタノール沈殿 ……………………………………… 38
    2. フェノール抽出 ……………………………………… 39
（2）培地の準備をする………………………………… 40
    1. 培地の作製（液体培地，寒天培地）……………… 40
    2. プレート培養 ………………………………………… 41
（3）大腸菌を培養する………………………………… 43
    1. 少量培養 ……………………………………………… 43
    2. 大量培養 ……………………………………………… 44
（4）プラスミド調整 ………………………………… 46
    1. アルカリプレップ …………………………………… 46
    2. 市販のキットを用いる ……………………………… 47
（5）制限酵素処理と電気泳動 ……………………… 49
    1. 制限酵素処理／電気泳動 …………………………… 49
（6）サブクローニング ……………………………… 52
    1. インサートDNAの準備 …………………………… 52
    2. ベクター調整 ………………………………………… 53
    3. ライゲーション ……………………………………… 54
    4. インサートDNAのチェック ……………………… 55
（7）形質転換（トランスフォーメーション）…… 57
    1. 塩化カルシウム法 …………………………………… 57
    2. エレクトロポレーション法 ………………………… 58
（8）発現解析を行う ………………………………… 59
    1. PCR法－①プライマーのデザイン ……………… 59
    2. PCR法－②PCR装置の設定 ……………………… 60
    3. ノーザンハイブリダイゼーション ………………… 61
    4. RNAaseプロテクションアッセイ ………………… 62
（9）塩基配列の決定（DNAシークエンス）…… 63
    1. 電気泳動……………………………………………… 63
    2. オートシークエンサーを用いた解析 ……………… 64
（10）データ解析とホモロジー検索 ……………… 66
    1. エレクトロフェログラム…………………………… 66
    2. データベース ………………………………………… 67

バイオ実験 英語でトライ！ Contents

Chapter 3　タンパク質実験をする ──────── 68
- （1）タンパク質の精製 ………………………………… 69
  - 1. タンパク質の抽出 ……………………………… 69
  - 2. タンパク質の分離精製（透析）………………… 70
- （2）精製タンパク質の分析 …………………………… 71
  - 1. タンパク質の定量 ……………………………… 71
  - 2. タンパク質の分子量測定 ……………………… 72
- （3）タンパク質の活性測定 …………………………… 73
  - 1. DNA結合活性測定 …………………………… 73
  - 2. DNA結合部位決定法 ………………………… 74
- （4）タンパク質の機能調節の解析 …………………… 75
  - 1. ウエスタンブロッティング …………………… 75
  - 2. 免疫沈降法 ……………………………………… 76
  - 3. Two-ハイブリッドシステム ………………… 77
- （5）タンパク質の分布の解析 ………………………… 78
  - 1. *in situ* での検出 ……………………………… 78
  - 2. *in vitro* での検出 …………………………… 79

Chapter 4　細胞培養をする ──────────── 82
- （1）培養器具／培養液の準備 ………………………… 84
  - 1. 滅菌と洗浄 ……………………………………… 84
  - 2. 培地をつくる …………………………………… 85
- （2）細胞の準備 ………………………………………… 87
  - 1. 凍結細胞の取り出し …………………………… 87
  - 2. 凍結細胞の融解 ………………………………… 88
- （3）細胞の観察 ………………………………………… 90
  - 1. 肉眼での観察 …………………………………… 90
  - 2. 顕微鏡での観察 ………………………………… 91
- （4）細胞の継代 ………………………………………… 93
  - 1. 前準備 …………………………………………… 93
  - 2. 継代 ……………………………………………… 94
- （5）細胞数の測定 ……………………………………… 96
  - 1. 細胞浮遊液の調整 ……………………………… 96
  - 2. 血球計算板での計測 …………………………… 97
- （6）細胞のクローニング ……………………………… 99
  - 1. コロニーの選定 ………………………………… 99
  - 2. 細胞の回収 ……………………………………… 100

バイオ実験 英語でトライ！ Contents

（7）細胞の保存 …………………………………………… 102
  1. 細胞懸濁液 ………………………………………… 102
  2. 凍結 ………………………………………………… 103
（8）培養実験を終えて ……………………………………… 105
  1. 培養器具の処理 …………………………………… 105
  2. 培養液の処理 ……………………………………… 106

SectionⅢ：4コマ漫画でシミュレーション～実験結果を報告しよう！

1　インサートの確認：制限酵素でDNAが切れない ……………………… 110
2　インサートの確認：制限酵素で予想外の断片が出る ………………… 111
3　形質転換の結果：コロニーが形成されない ……………………………… 112
4　形質転換の結果：セルフライゲーションのコロニーのみ形成 ………… 113
5　PCR産物の確認：増幅バンドが検出されない …………………………… 114
6　PCR産物の確認：非特異的バンドが多い ………………………………… 115
7　ノーザンブロッティングの結果：バンドが検出されない ……………… 116
8　ノーザンブロッティングの結果：バックグラウンドが高い …………… 117
9　ウエスタンブロッティングの結果：バンドが検出されない …………… 118
10　ウエスタンブロッティングの結果：バックグラウンドが高い ………… 119

索引：キーワードで探す英語表現 …………………………………………… 120

コラム

アメリカ式ラボファッションって？	29	開かずの扉	80
気長に待つのがお国柄？	29	コンタミについて（日本とボストンの培養環境のちがい）	86
ちょっとお騒がせな火災感知器	34	アイソトープも一般の実験室で？！	92
計画通りに進まないもの…？	45	テクニシャンとの関係	95
アメリカのラボは骨董品の宝庫？	45	スペイン語も？	95
Budget… ご利用は計画的に	48	ラボミーティング	98
実験台も背が高い！	51	慣れは恐い	101
勇気を出して	65	論文の準備… でも進まない	104
長蛇のレジ！その理由は…	65	ラボの廊下	107
Surprising Party（Birthday）	77		
切り換えも大切？	80		

Section I

イラストでわかる実験室での英語表現

Chapter1 意外に知らない実験機器の名前 … 12

Chapter2 目で見てわかるニュアンスのちがい … 18

I●1 意外に知らない実験機器の名前

❶ 実験台の周辺

・自分の実験台の上には何がありますか？ 毎日使用する器具の名称を紹介します．

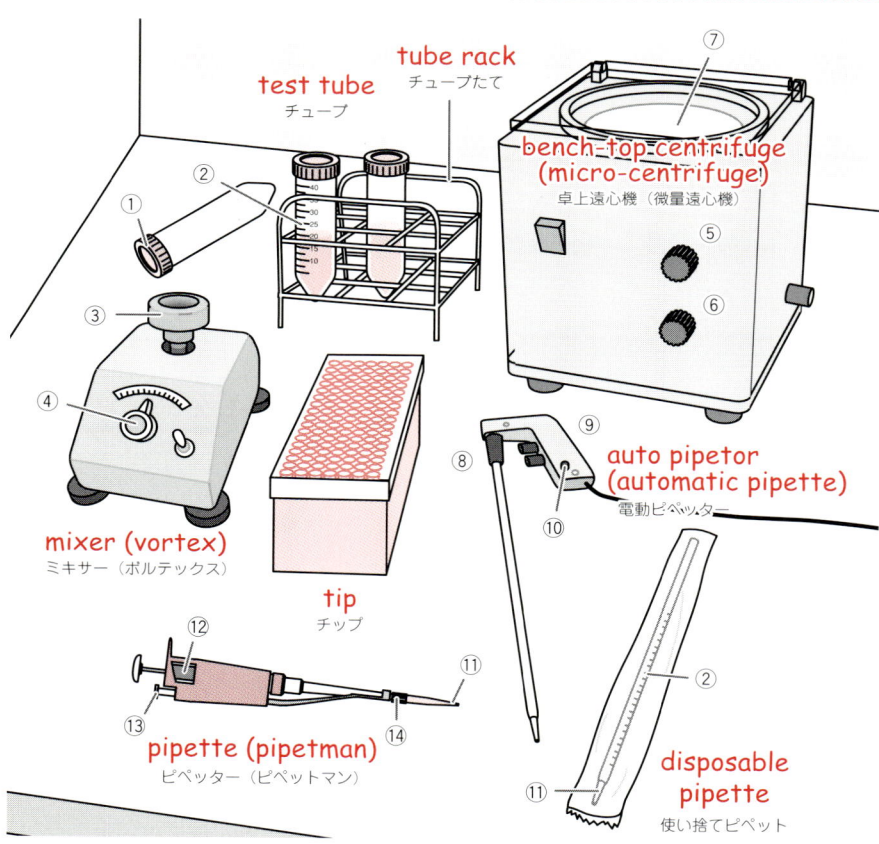

① cap キャップ　② scale 目盛り　③ vibrator unit/cup 振動板　④ vibration control 振動調節つまみ
⑤ timer タイマー　⑥ rotation control 回転数調節つまみ　⑦ rotor ローター
⑧ connector コネクター　⑨ grip 取っ手　⑩ speed control 速度調節つまみ
⑪ ejection site 液体流出口　⑫ scale adjustment screw 目盛り調節ねじ
⑬ tip replacement button チップ交換ボタン　⑭ pipette connector ピペッターとの接続部分

I●1 意外に知らない実験機器の名前

2 計量台の周辺

・試薬を調整する際に必要な器具です．
ガラス製品が多いため，破損に気をつけて使用しましょう．

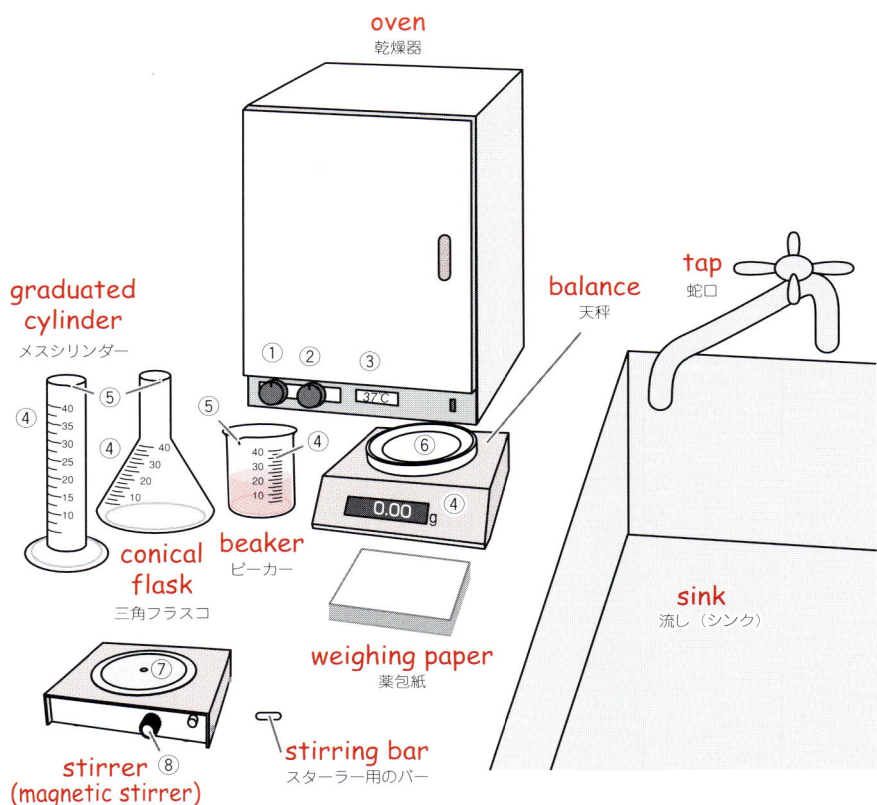

① temperature control 温度調節つまみ　② timer タイマー　③ temperature indicator 温度表示窓
④ scale 目盛り　⑤ lip 注ぎ口　⑥ balance dish 天秤皿
⑦ beaker pad/beaker plate ビーカー設置板　⑧ rotation control 回転数調節つまみ

I ●1 意外に知らない実験機器の名前

❸ DNA実験の実験台周辺

・DNAを扱う実験台周辺には，DNA分解酵素が混入しないようにしましょう．

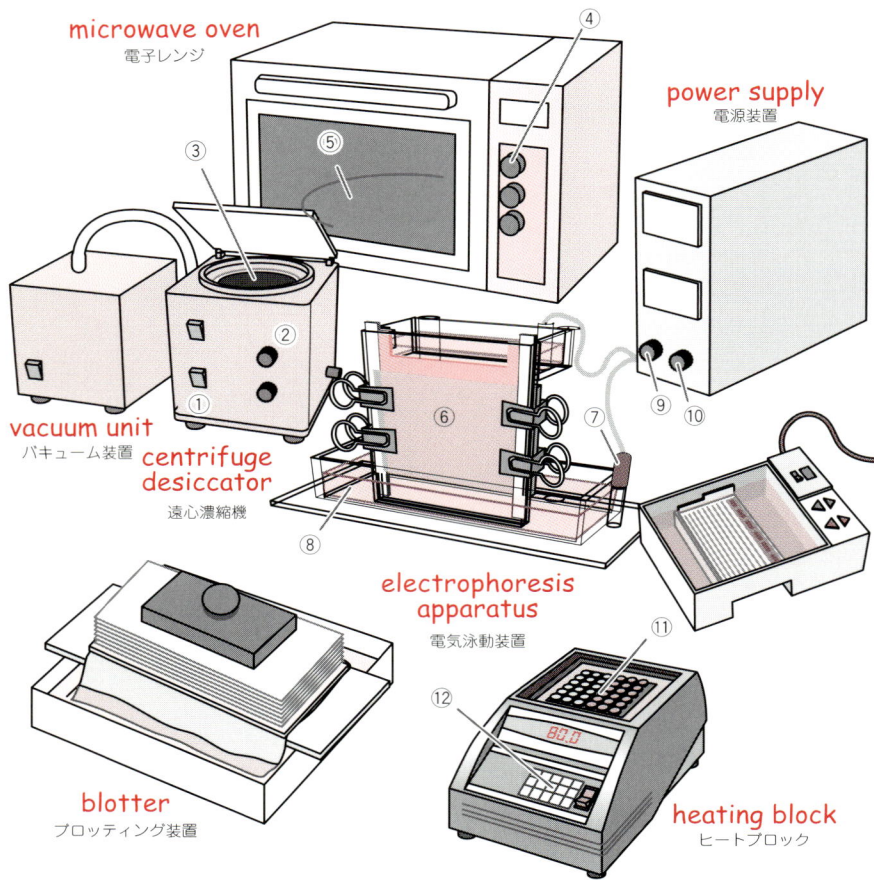

microwave oven 電子レンジ
power supply 電源装置
vacuum unit バキューム装置
centrifuge desiccator 遠心濃縮機
electrophoresis apparatus 電気泳動装置
blotter ブロッティング装置
heating block ヒートブロック

① vaccum switch バキュームスイッチ　② rotation control 回転数調節つまみ　③ rotor ローター
④ timer タイマー　⑤ rotator 回転台　⑥ gel cassette ゲル作製枠
⑦ electric cord connector 電極線接続部　⑧ electrophoresis tank 電気泳動槽
⑨ voltage control 電圧調節つまみ　⑩ mA control 電流調節つまみ　⑪ block ブロック
⑫ temperature control 温度調節ボタン

I●1 意外に知らない実験機器の名前
❹ 暗 室

・自動現像機は安全光のもとで使用します．紫外線により眼や皮膚に影響を受けないように注意します．また，RIの被曝は最小限に抑えましょう．

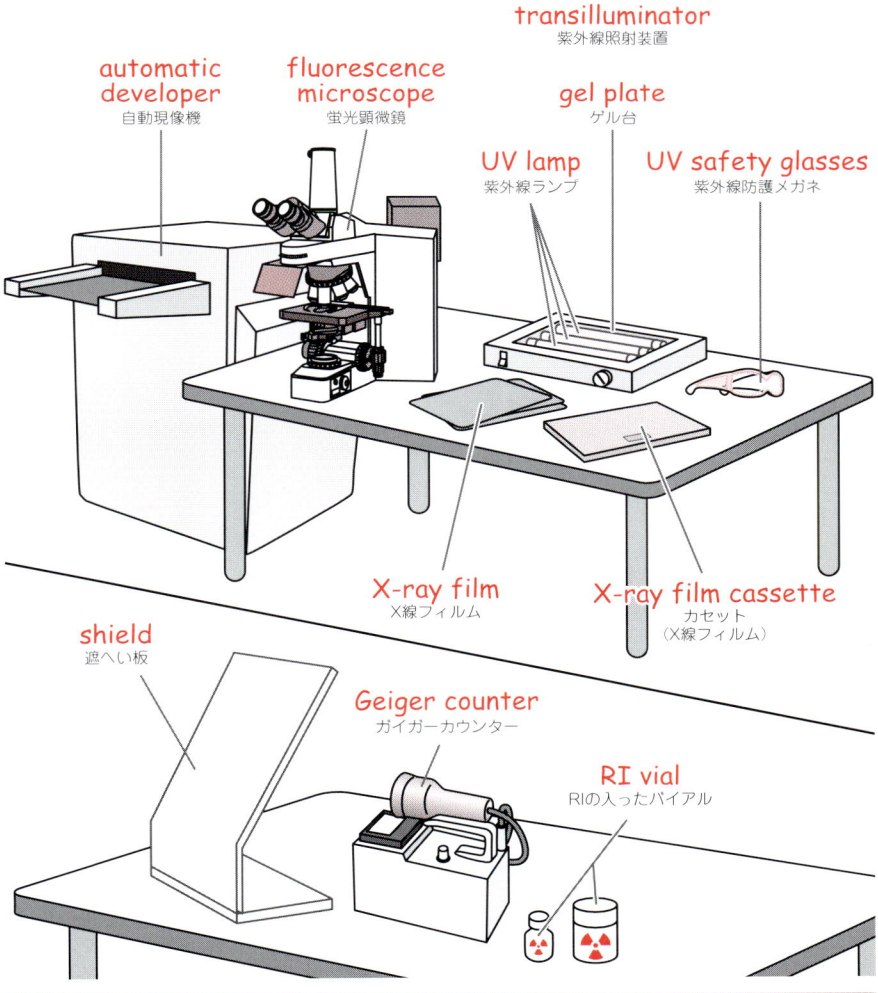

Section I：イラストでわかる実験室での英語表現

I・1 意外に知らない実験機器の名前

5 細胞培養室

・無菌操作のための設備です．各研究室でのルールを守り，汚染（contamination）の起らないよう気をつけましょう．

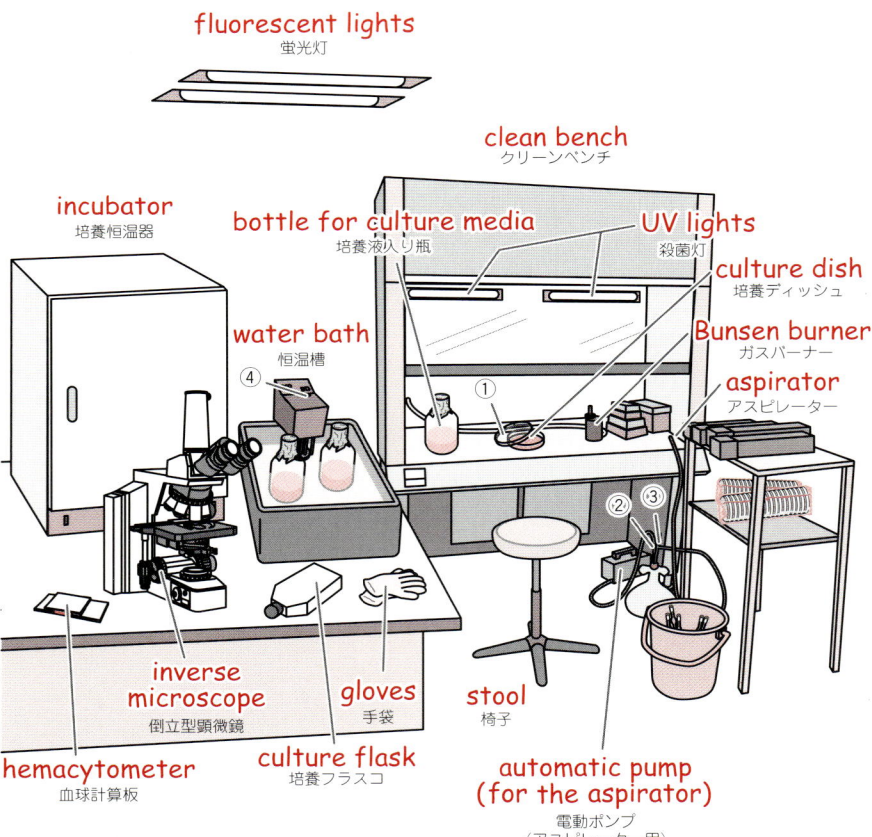

① cover 蓋 ② connector to aspirator 吸引管接続口
③ connector to Pasteur pipette パスツールピペット接続口
④ temperature control 温度調節つまみ

I・1 意外に知らない実験機器の名前

6 クリーンベンチの周辺

・大腸菌を扱った実験を行う部屋です．
使用済みの器具類はオートクレーブで滅菌します．

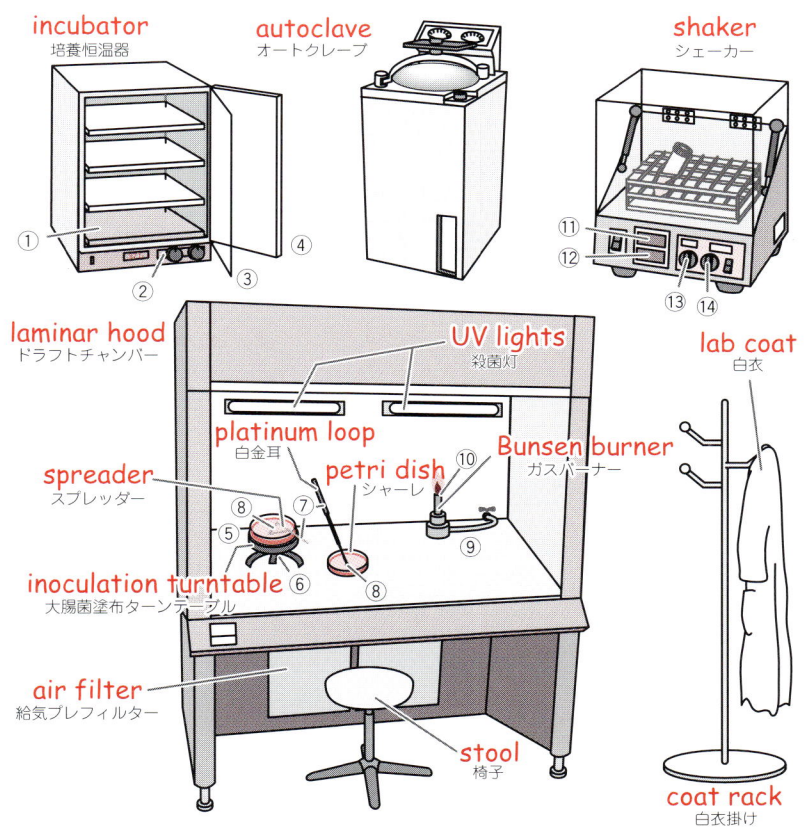

① water reservoir 水を入れるトレイ
② temperature and CO_2 concentration indicator 温度，炭酸濃度調節パネル
③ inner door 内扉　④ outer door 外扉　⑤ petri dish table シャーレ設置面　⑥ rotation axis 回転軸
⑦ grip 持ち手　⑧ spreading area 延ばす所　⑨ gas tube ガス管　⑩ gas tap ガス流出口
⑪ rotation speed indicator 回転数表示窓　⑫ temperature indicator 温度表示窓
⑬ rotation control 回転数調節つまみ　⑭ temperature control 温度調節つまみ

I ● 2 目で見てわかるニュアンスのちがい

❶ 水の種類

・実験に使用する水にもいろいろあります．用途によって使い分けましょう．

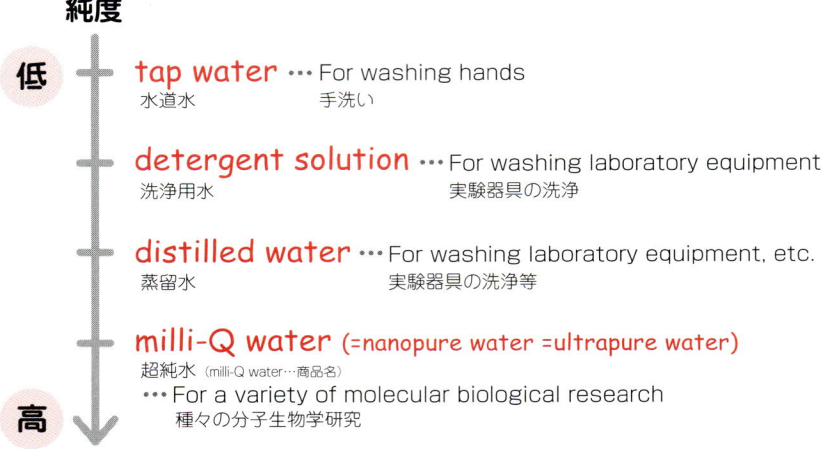

純度

低 ↑
- **tap water** … For washing hands
  水道水　　　　手洗い
- **detergent solution** … For washing laboratory equipment
  洗浄用水　　　　　　実験器具の洗浄
- **distilled water** … For washing laboratory equipment, etc.
  蒸留水　　　　　　実験器具の洗浄等
- **milli-Q water** (=nanopure water =ultrapure water)
  超純水　(milli-Q water…商品名)
  … For a variety of molecular biological research
  　　種々の分子生物学研究

高 ↓

使用目的でわけると…

●For molecular biological research 分子生物学研究用

- **ddH₂O-distilled deionized water** (=double distilled water)
  脱イオン化蒸留水
  … For a variety of molecular biological research
  　　種々の分子生物学研究
- **DEPC-treated water** … For RNA experiments
  DEPC添加水　　　　　　　　RNA実験

●For cell culture 細胞培養用

- **autoclaved water** … For DNA experiments and cell culture
  オートクレーブ水　　　　DNA実験，細胞培養
- **filtered, sterilized water** … For cell culture
  濾過滅菌水　　　　　　　　細胞培養

I•2 目で見てわかるニュアンスのちがい

❷ 混ぜる操作

・日本語で「混ぜる」と一口に言っても実際にはさまざまな方法，ニュアンスのちがいがあります．ここでは代表的なものを紹介してみましょう．

混ぜる力

弱

- **tapping**
  指ではじく

- **flipping upside down**
  転倒混和

- **pipetting**
  ピペッティング

- **shaking vigorously**
  上下に強くふる

- **mixing on a stirrer**
  スターラーで混ぜる

- **vibrating on a vortex**
  ボルテックスをかける

強

Section I：イラストでわかる実験室での英語表現

I ● 2 目で見てわかるニュアンスのちがい

❸ 泳動ゲルの種類と分子量

・電気泳動するサンプルの種類と分子量によってゲルの種類が変わってきます．目的に応じて使い分けましょう．

● SDS-PAGE SDS-PAGE用ゲル
(sodium dodecylsulfate-polyacrylamide gel electrophoresis)

・・・Determination of protein purity and estimation of the molecular weight
タンパク質の純度の検定，分子量の推定

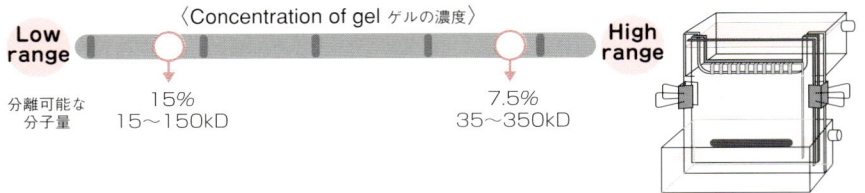

Low range / High range
〈Concentration of gel ゲルの濃度〉
分離可能な分子量
15% 15〜150kD
7.5% 35〜350kD

● Agarose gel アガロースゲル

・・・Determination of DNA purity and estimation of the molecular weight
DNAの純度の検定，分子量の推定

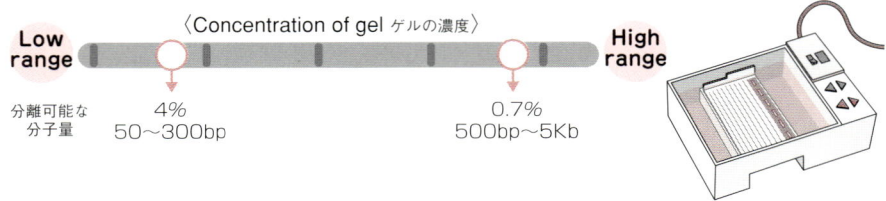

Low range / High range
〈Concentration of gel ゲルの濃度〉
分離可能な分子量
4% 50〜300bp
0.7% 500bp〜5Kb

● Denaturing agarose gel (agarose formaldehyde gel)
変性アガロースゲル（ホルムアルデヒド変性ゲル）

・・・Determination of RNA (mRNA) purity and estimation of the molecular weight
RNA(mRNA)の純度の検定，分子量の推定

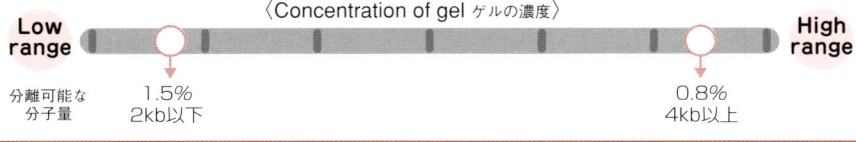

Low range / High range
〈Concentration of gel ゲルの濃度〉
分離可能な分子量
1.5% 2kb以下
0.8% 4kb以上

I・2 目で見てわかるニュアンスのちがい
4 泳動ゲルおよびメンブレンの染色法

・染色法によって検出感度が異なってきます．ゲルやメンブレンの種類によって染色法を選びましょう．

● Gels ゲル

Nucleic acid 核酸

- SYBR-Green staining
  SYBR-Green染色
- SYBR-Gold staining
  SYBR-Gold染色
- Ethidium bromide staining
  エチジウムブロマイド染色

High 〈Detection sensitivity〉 Low
検出感度

- Silver staining
  銀染色
- CBB(Coomassie Brilliant Blue G-250) staining
  CBB 染色

Protein タンパク質

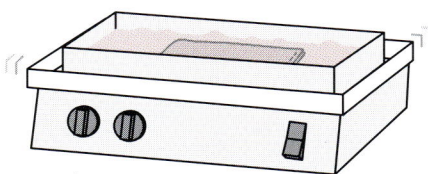

- Destaining is difficult.
  脱色困難
- Destaining is easy.
  脱色容易

● Membranes メンブレン

High 〈Detection sensitivity〉 Low
検出感度

- Colloidal Gold staining
  Colloidal Gold染色
- CBB staining
  CBB染色
- Ponceau S staining
  Ponceau S染色

I ●2 目で見てわかるニュアンスのちがい
❺ 実験材料の保存

・実験材料の保存は重要です．材料によって保存温度が異なるので注意しましょう．

●refrigerator (4℃) 冷蔵庫 (4℃)

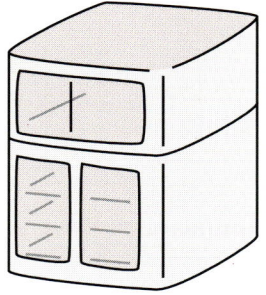

··· For liquid media and antibodies
液体培地，抗体

●freezer 冷凍庫 (−20℃, −80℃)

●−20℃ ··· For restriction enzymes, antibodies, DNA samples
制限酵素，抗体，DNAサンプル

●−80℃ ··· For frozen cells (for short periods of time), RNA samples
凍結細胞（短期間），RNAサンプル

●programmed freezer
プログラムフリーザー

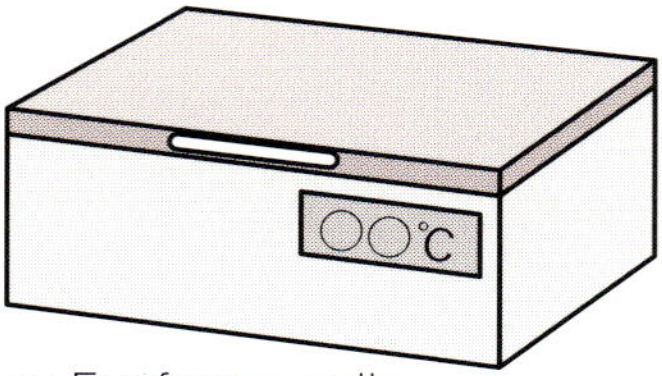

··· For frozen cells
凍結細胞

●liquid nitrogen tank
液体窒素タンク

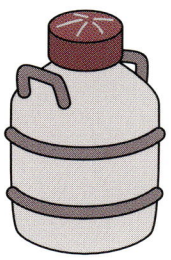

··· For frozen cells (for long periods of time)
凍結細胞（長期間）

Section II

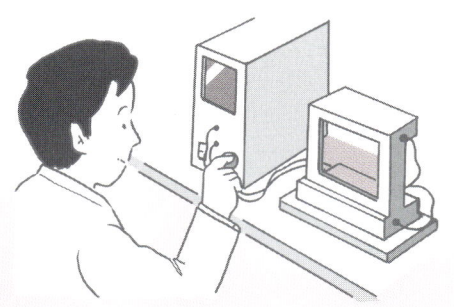

Let's speak!
~実験中の英会話

Chapter 1 実験の準備・・・・・・・・・・・・・・・・・・・・24

Chapter 2 DNA実験をする・・・・・・・・・・・・・・36

Chapter 3 タンパク質実験をする・・・・・・・・・68

Chapter 4 細胞培養をする・・・・・・・・・・・・・・82

Chapter 1
実験の準備

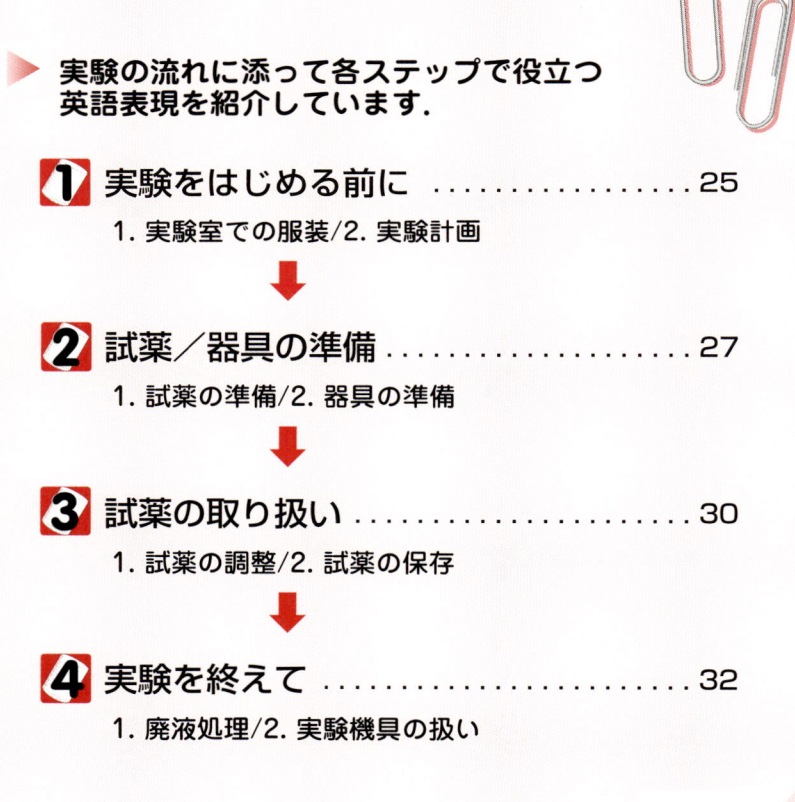

▶ 実験の流れに添って各ステップで役立つ英語表現を紹介しています．

❶ 実験をはじめる前に 25
　1. 実験室での服装／2. 実験計画

❷ 試薬／器具の準備 27
　1. 試薬の準備／2. 器具の準備

❸ 試薬の取り扱い 30
　1. 試薬の調整／2. 試薬の保存

❹ 実験を終えて 32
　1. 廃液処理／2. 実験機具の扱い

II●1 実験の準備

❶ 実験をはじめる前に

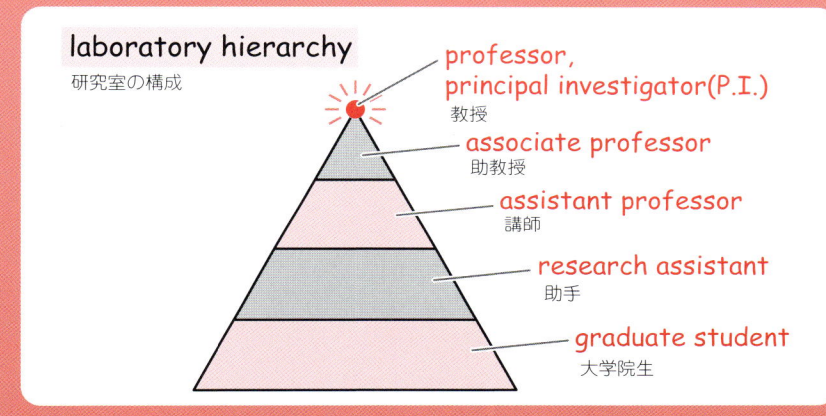

laboratory hierarchy
研究室の構成
- professor, principal investigator(P.I.) 教授
- associate professor 助教授
- assistant professor 講師
- research assistant 助手
- graduate student 大学院生

1. 実験室での服装
…What to wear in the laboratory

● **What are the proper** clothes to wear in a laboratory?
● 実験室での正しい服装とはどのようなものでしょうか？

　🔖 その他の表現法
　◆ **What kind of** clothes should you wear in a laboratory?

● Do you need to roll up your sleeves when you wear a lab coat?
● 白衣を着る場合，袖まくりしたほうがよいのでしょうか，しないほうがよいのでしょうか？

● What are the proper clothes to wear when handling nucleic acid
● DNA，RNAなどの核酸を使う場合にどのような服装を

such as DNA or RNA? | したらよいのでしょう？

🌑 What are the proper clothes for cell culture? | 🌑 細胞培養をする場合にはどのような服装をしたらよいのでしょうか？

🌑 What are the proper clothes when handling radioisotopes? | 🌑 アイソトープを扱うときにはどのような服装をしたらよいのでしょうか？

🔴 You need to wear a special coat for radioisotopes over your lab coat. | 🔴 それには，白衣の上にアイソトープ用の着衣という格好をしてください．

2. 実験計画
…Experimental plan

🌑 **How do you collect** references before performing experiments? | 🌑 実験をする前にどのようにして文献を集めればよいですか？

　🍪 その他の表現法
　◆ *Where do you find references before performing experiments?*
　◆ *How do you search for references before performing experiments?*

🌑 How do you establish a plan for performing the experiment? | 🌑 この実験を行うにあたってどのような実験計画をたてますか？

II・1 実験の準備

2 試薬／器具の準備

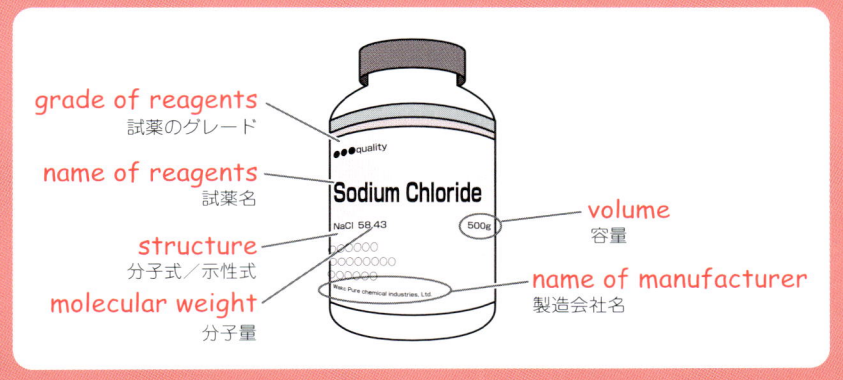

1. 試薬の準備
…Preparation of reagents

● **What kinds of reagent do you need** to perform this experiment?

　🐾 その他の表現法
　◆ *What kinds of reagent are required to perform this experiment?*

● この実験を行うにあたってどのような試薬が必要になりますか？

● Check the protocol to confirm the catalogue number, and, if necessary, the lot number of the reagent.

● プロトコールを確認して，試薬のカタログ番号や場合によってはロット番号を合わせます．

● Are all the reagents stored in a common (interdepartmental) reagent cabinet?

● 試薬は全て共通の薬品棚にありますか？

Section II：Let's speak! 〜実験中の英会話

🦐 No, reagents designated as poisonous or deleterious substances are stored separately from other reagents.

🦐 いいえ，毒物および劇物扱いのものは一般の試薬と別に保管されています．

2. 器具の準備
…Preparation of laboratory equipment

🦐 Which experimental machines (apparatus/equipment) do you need **to reserve** to perform this experiment?

🦐 この実験を行うにあたってどの実験機器を予約する必要がありますか？

🐛 その他の表現法
◆ Which experimental machines (apparatus/equipment) do you need <u>to put your name down for</u> to perform this experiment?

🦐 Please make a reservation for an autosequencer in the common (interdepartmental) laboratory.

🦐 共通実験室にあるオートシークエンサーを予約してください．

🦐 **Can you use** the automatic processor **immediately after switching it on**?

🦐 自動現像機は，スイッチを入れるだけですぐ使えますか？

🐛 その他の表現法
◆ <u>Can you use</u> the automatic processor <u>straight after you switch it on</u>?

🐟 In winter, you should switch it on at least 20 minutes before you want to use it.

🐟 冬場は最低，使用20分前にはスイッチを入れるようにしてください．

Column　アメリカ式ラボファッションって？

　アメリカのラボでは研究者の服装はまちまちでした．私が留学していたラボは多国籍ラボでしたが，白衣を一番よく着ていた人種は日本人だったと思います．ただし，われわれの所属していたチームの日本人ボスはいつもオペ着姿でした．動物を扱った実験が盛んに行われていたラボだったこともあるかと思います．P2やアイソトープ使用の場合は当然皆着替えていましたが，それ以外の区域ではそれぞれの好みの服装でした．

Column　気長に待つのがお国柄？

　アメリカのラボといってもいろいろなラボがあると思いますが，私の行っていたラボの様子や諸先輩方が留学されていたラボでの話を総合すると，実験機器は日本のラボの方がずっとマシです．よく壊れますし，壊れたときの対応がお国柄でとてものんびりです．休みの前等は特にはっきりしています．日本だとお得意様ということで無理を押してきていただけることもありますが，アメリカではまずあり得ません．実験とは関係ありませんが銀行窓口の対応も日本人には戸惑うことの１つです．ときには気長に待つことも必要だということを感じました．

Ⅱ●1 実験の準備

❸ 試薬の取り扱い

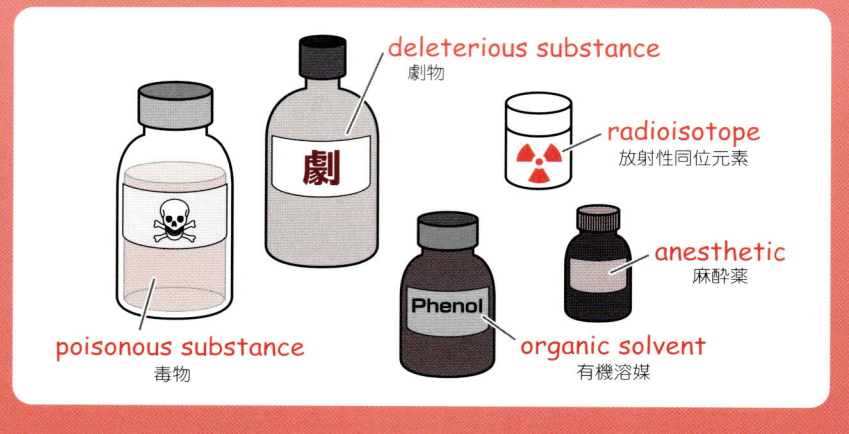

1. 試薬の調整
…Preparation of reagents

🎃 What kind of water do you use to prepare reagents?

🎃 試薬の調整にどんな水を使用しますか？

🍎 In this case, you use autoclaved milli-Q water.

🍎 今回は超純水をオートクレーブしたものを使用します．

🎃 What should I do to adjust the pH of this reagent to 7.2?

🎃 この試薬のpHを7.2に合わせたいのですが？

🍎 You can **adjust** the pH with an acid or alkali **using a pH meter**.

🍎 滴定用の酸，またはアルカリを使って，pHメーターで合わせてください．

その他の表現法
◆You can **match** the pH **to** the required value with an acid or alkali *using a pH meter*.

2. 試薬の保存
…Storage of reagents

🍬Where should you store this reagent?

この試薬はどこに保存すればよろしいでしょうか？

🍬**Shield it from the light** and store it in a refrigerator at 4℃.

遮光して4℃の冷蔵庫に保存してください．

その他の表現法
◆*Protect it from the light* and store it in a refrigerator at 4℃.

🍬What is the **expiration date** of this reagent?

この試薬の使用期限はいつまでですか？

その他の表現法
◆*How long can you use* this reagent *for*?

🍬It should be printed on the label on the bottle.

試薬瓶のラベルに記載されてあるはずです．

Ⅱ●1 実験の準備

4 実験を終えて

1. 廃液処理
…*(Laboratory) waste disposal*

🍴 Can you **throw** this waste water **away in the sink**?

🍴 この廃液はシンクに流してもよろしいですか？

その他の表現法
◆ Can you **dispose of** this waste water **in the sink**?

🍴 No, you should add hypochlorous acid and dispose of it as hazardous industrial waste.

🍴 いいえ，次亜塩素酸を加えて産業廃棄物として廃棄します．

🐟 **How should you dispose of** the waste from the P2-level laboratory?

🐠 P2 実験室で出た廃液はどのように処理しますか？

🐡 その他の表現法
◆ *How should you handle* the waste from the P2-level laboratory?

🐟 In principle, it should be autoclaved, and then a waste management company should be asked to collect and dispose of it.

🐠 原則としてオートクレーブしてから業者に廃棄を依頼します．

2. 実験機具の扱い
…Handling laboratory equipment

🐟 **How should you keep a record** after you use the ultra-centrifuge in the common (interdepartmental) laboratory?

🐠 共通室にある超遠心機を使用後，記録はどうしますか？

🐡 その他の表現法
◆ *What records should you keep* after you use the ultra-centrifuge in the common (interdepartmental) laboratory?

🐟 You should record the time you finished using it and the meter count.

🐠 終了時の時間と，メーター数を記入しておきます．

🐟 What should you do with the electrophoresis tank when you've finished with it?

🐟 使用後の電気泳動槽はどうしますか？

🐟 You should wash it lightly with tap water, pour distilled water over it and leave it to dry.

🐟 水道水で軽く洗浄し蒸留水を流したあと，乾燥させておきます．

Column ちょっとお騒がせな火災感知器

　アメリカの火災報知器の警告音は半端ではありませんがそれよりも，どんなに誤報の可能性が高くても，どんなに外が吹雪の日でも，いったん外へ出て，警備員が確認するまで中に入れないのです．ある冬の日の朝，培養室で細胞の継代操作の途中だったとき，突然けたたましいサイレンが鳴りだしました．日本の感覚で誤報だろうと思い，大事な操作中だったこともありそのまま作業を続けていました．すると警備員が培養室に飛び込んできて「何してる！　早く外へ出ろ」と言って，つまみだされました．そこにはラボの大勢が寒そうに立っていました．結局誤報だったのですが，日本だとちょっと考えられないことでした（本当は，日本でもちゃんと避難しないといけないのですが）．大事な細胞は駄目になってしまいましたが，本当の火災で，異国で果てるのも辛いですので，やはりすぐに逃げた方がよいでしょう．（後になって，アメリカの火災感知器は日本の数倍鳴りやすいことがわかりましたが），このように，実験には予期せぬ出来事が起こる可能性があります．

MEMO

Chapter 2
DNA実験をする

▶ 実験の流れに添って各ステップで役立つ英語表現を紹介しています．

❶ DNAを抽出する 38
　　1. エタノール沈殿／2. フェノール抽出

❷ 培地の準備をする 40
　　1. 培地の作製（液体培地，寒天培地）／2. プレート培養

❸ 大腸菌を培養する 43
　　1. 少量培養／2. 大量培養

❹ プラスミド調整 46
　　1. アルカリプレップ／2. 市販のキットを用いる

❺ 制限酵素処理と電気泳動 49
　　1. 制限酵素処理／電気泳動

6 サブクローニング 52
1. インサートDNAの準備／2. ベクター調整
3. ライゲーション／4. インサートDNAのチェック

7 形質転換（トランスフォーメーション）.... 57
1. 塩化カルシウム法／2. エレクトロポレーション法

8 発現解析を行う 59
1. PCR法－①プライマーのデザイン
2. PCR法－②PCR装置の設定
3. ノーザンハイブリダイゼーション
4. RNAase プロテクションアッセイ

9 塩基配列の決定（DNAシークエンス）... 63
1. 電気泳動／2. オートシークエンサーを用いた解析

10 データ解析とホモロジー検索 66
1. エレクトロフェログラム／2. データベース

II•2 DNA実験をする

II・2 DNA実験をする

❶ DNAを抽出する

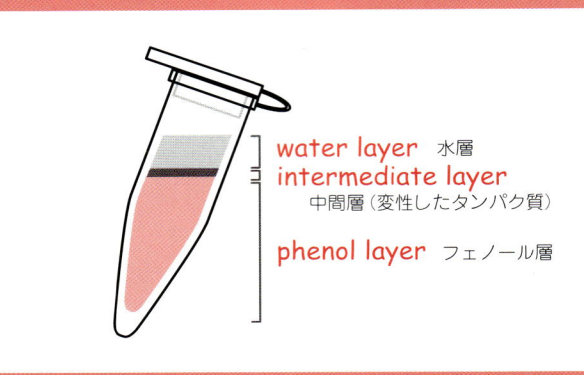

water layer　水層
intermediate layer
　中間層（変性したタンパク質）
phenol layer　フェノール層

1. エタノール沈殿
…Ethanol precipitation

🍬 **Under what conditions do you** precipitate ethanol?

🍬 エタノール沈殿はどんな条件で行いますか？

🍬 You incubate the tube at -80℃ for more than 30 minutes and then centrifuge it.

🍬 －80℃で30分以上おいてから遠心をします．

🍬 **What's the next step after centrifugation?**

その他の表現法
◆ *What do you do after* you (do something)?
◆ *What do you do once* you (have done something)?

🍬 遠心した後はどのようにしますか？

- You wash the pellet in 75% ethanol and then evaporate the ethanol.

75％エタノールでペレットを洗浄し乾燥させます.

2. フェノール抽出
…Phenol extract

- What's the ratio of phenol to nucleic acid solution when they are mixed?

フェノールと核酸の溶液はどのような割り合いで混合しますか？

- **The ratio is 1:1**.

1:1で混合します．

One Point
◆どう読めばいいのかわかりますか？
The ratio is 1:1はthe ratio is one to oneと読みます．

- How should you deal with the white layer between the phenol and nucleic acid solution?

中間の白色層はどのように扱いますか？

- You should avoid aspirating the white layer into the nucleic acid solution.

核酸溶液の中に吸引しないように注意します．

II●2 DNA実験をする

❷ 培地の準備をする

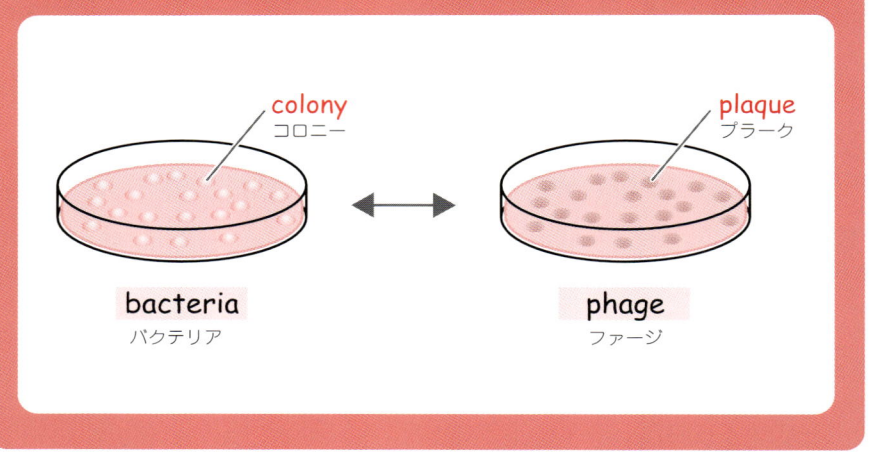

- colony コロニー
- bacteria バクテリア
- plaque プラーク
- phage ファージ

1. 培地の作製（液体培地，寒天培地）
…Preparation of culture media (liquid media and agar media)

🍀 What's the difference **between** liquid media and agar media?

その他の表現法
◆ *In what way do* liquid media and agar media *differ*?

One Point
◆ 3つのものを比べるときは次のようにいいます．
What's the difference between A, B and C?
＝What's the difference among A, B and C?

🍀 液体培地と寒天培地の違いはなんですか？

🐡 The difference is whether the media contains agar or not.

🟡 アガー（寒天）を含むか，含まないかです．

🟢 Under what conditions is the media stored?

🟢 培地の保存はどのようにしますか？

🟢/その他の表現法
◆ How do you store the media?
◆ What conditions do you store the media under?（より口語的）

🐡 You add antibiotics and store at a temperature of 4℃.

🟡 抗生物質を加えた後，4℃で保存します．

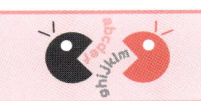

2. プレート培養
…Plate culture

🟢 How do you handle plates stored at 4℃?

🟢 4℃で保存してあるプレートを最初どうしますか？

🟢/その他の表現法
◆ What's the first thing you do with plates stored at 4℃?

🐡 You incubate them in an incubator at 37℃ for 30 minutes.

🟡 37℃恒温器の中に，30分ほど入れておきます．

🐡/その他の表現法
◆ You place them in an incubator and incubate at 37℃ for 30 minutes.

🐛 How do you **amplify** *E. coli* on the agar media?

> 🐛 その他の表現法
>
> ◆大腸菌等を増やす場合には，amplifyが多く使用されますが次のような表現もあります．
>
> *How do you **grow（multiply / culture）** E. coli on the agar media?*

🐛 寒天培地でどのように大腸菌を増やしますか？

💬 You use a spreader on an inoculation turntable.

💬 大腸菌塗布ターンテーブルの上でスプレッダーを用いて行います．

🐛 How do you incubate the plate after spreading the *E. coli*?

🐛 大腸菌塗布後のプレートをどのように保存しますか？

💬 You incubate the plate **upside down** in the incubator.

> 🐛 One Point
>
> ◆上下反対にしてものを置く場合などに用います．

> 🐛 その他の表現法
>
> ◆*You **invert** the plate in the incubator.*
>
> ◆*You **turn** the plate **upside down** in the incubator.*

💬 上下反対にして，培養恒温器の中でインキュベートします．

II•2 DNA実験をする

3 大腸菌を培養する

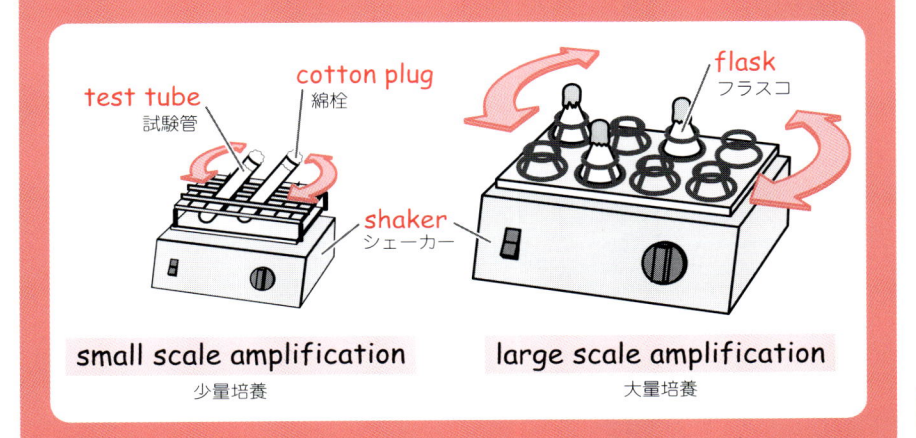

test tube 試験管
cotton plug 綿栓
shaker シェーカー
flask フラスコ

small scale amplification
少量培養

large scale amplification
大量培養

1. 少量培養
…Small-scale amplification

● **How do you** incubate liquid media?

● 液体培養はどのような条件で行いますか？

その他の表現法
◆ <u>Under what conditions do you</u> incubate liquid media?
◆ <u>What conditions do you</u> incubate liquid media under?

● (Liquid media are incubated) in a shaker at 37℃.

● 37℃のシェーカー内で培養します．

その他の表現法
◆ *They're incubated in a shaker at 37℃.*

Section II：Let's speak! 〜実験中の英会話

● What do you need to be careful about when incubating liquid media?

●液体培養ではどのようなことに気をつけますか？

●**You should take care not to over-incubate** the liquid media.

●あまり長くインキュベーションし過ぎないように気をつけます．

🐾その他の表現法
◆ <u>You should be careful not to</u> incubate the liquid media <u>for too long</u>.
◆ <u>You need to take care not to</u> incubate the liquid media <u>too much</u>.

2. 大量培養
…Large-scale amplification

● When should you add an antibiotic to liquid medium?

●液体培地の中に入れる抗生物質はいつ加えますか？

● It should be added after the liquid medium has been placed in a flask and autoclaved.

●液体培地をフラスコに入れ，オートクレーブした後に加えます．

● Should you always use the same antibiotic in liquid media?

●液体培地に加える抗生物質はいつも同じですか？

● The types and concentrations of antibiotics **vary depending on**

●形質転換に用いられたベクターの種類によって抗

what kind of vector is used for transformation.

🔸その他の表現法
◆ The types and concentrations of antibiotics <u>differ according to the type of</u> vector that is used for transformation.

生物質の種類や濃度が変わります．

Column 計画通りに進まないもの…？

　いざアメリカへ留学して，自分の新しいテーマを見つけてはじめから自分が実験計画を綿密に立てていくことは一般的な流れですが，必ずしもこのような状況ばかりではありません．ときには志なかばにして帰国した研究者のテーマの後を引き継ぐ場合もあります．この場合は実験の流れはある程度決まっているのですが，自分自身で立てた計画でないため戸惑いがあります．十分引き継ぎをしたつもりでも，当事者が故郷（ヨーロッパ等）に帰った後で，試薬の位置がわからなかったりして連絡を取ることもあります．このときその当事者も帰国直後でバタバタしておりなかなか連絡が取れず実験が進まないこともあります．このような例もありますので実験計画を立てるときはいろんな状況を考えておいた方がよいと思います．

Column アメリカのラボは骨董品の宝庫？

　日本のラボでも骨董品寸前の乾燥機がまだ稼働してたりすることはよくありますが，アメリカのラボで最初に機器の説明をしてもらっていたときにオートクレーブの機器を見てその古さに驚きました．まるで陶器を焼く窯のような印象を受けました．この機器を使っていろんな実験をしましたが幸い問題はありませんでした．外観は近代的な立派な建物でも，中には案外古い機器が置いてあるかもしれません．

Section Ⅱ：Let's speak! ～実験中の英会話

II●2 DNA実験をする

❹ プラスミド調整

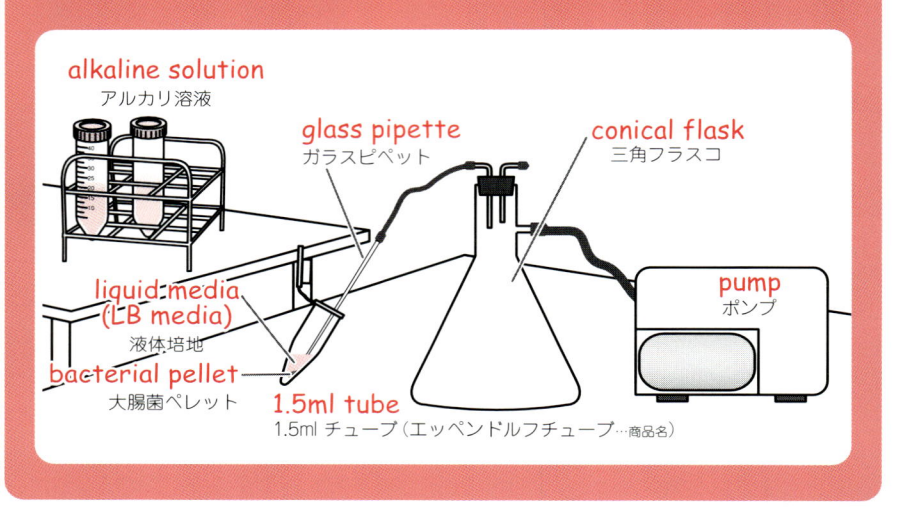

- **alkaline solution** アルカリ溶液
- **glass pipette** ガラスピペット
- **conical flask** 三角フラスコ
- **liquid media (LB media)** 液体培地
- **bacterial pellet** 大腸菌ペレット
- **1.5ml tube** 1.5ml チューブ（エッペンドルフチューブ…商品名）
- **pump** ポンプ

1. アルカリプレップ
…Alkaline prep

🍫 What do you have to be careful about when you add alkaline solution to the bacterial pellet?

🍫 集菌したペレットにアルカリ溶液を加えるとき，どのようなことに気をつけますか？

🍎 You should take care not to exceed the **reaction time (incubation).**

🍎 反応時間を超えないように気をつけます．

その他の表現法
◆ You should take care not to exceed the <u>time needed for a reaction (an incubation)</u>.

🍋 **How do you handle** RNA contamination?

🍋 RNAのコンタミにはどのように対処しますか？

🐛 その他の表現法
◆ <u>*What do you do about*</u> *RNA contamination?*

🌙 You add RNAase in advance.

🌙 事前にRNAaseを加えます．

2. 市販のキットを用いる
…Using commercial kits

🍋 What kinds of plasmid preparation kits are available in the market?

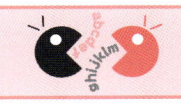

🍋 プラスミド調整に用いられる市販のキットにはどのようなものがありますか？

🐛 その他の表現法
◆ *What types of plasmid preparation kits are available in the market?*
◆ *What sort of plasmid preparation kits are available in the market?*
◆ *Is there a variety of plasmid preparation kits available in the market?*
（口語）

🌙 There is one kit where **DNA is adsorbed to a column membrane** and then eluted.

🌙 カラムのメンブレンにDNAを吸着させ，最後に抽出してくるタイプのものがあります．

🐛 その他の表現法
◆ *Yes, there are several types. For example, there is one kit where* <u>*DNA*</u>

Section Ⅱ：Let's speak! 〜実験中の英会話

is adsorbed to a column membrane and then eluted.（口語）

● **What is the advantage of using** a commercially available plasmid prep kit?

●市販のキットを用いる利点は何ですか？

🔖その他の表現法
◆ *What benefit is there to using a commercially available plasmid prep kit?*

● It gives higher purity DNA in a short period of time.

●短時間でより純度の高いDNAを回収できることです．

Column: Budget…ご利用は計画的に

　留学先は，Bossが結構grantを取っていて，実験材料は自由に購入できると思っていました．しかし，実際には研究者1人あたりの，budgetが決められていました．これは，研究費が湯水のようにあるのではなく，しっかりした実験計画を基に材料を購入するようにという意味だったと思っています．使用する材料には，期限があるものもあり（抗体等），その場その場でものを購入するのではなく，ある程度最初の段階で計画を立てることの重要性を感じました．しっかりした計画で，見込みがあると認められれば，必要な研究費は十分に支給されると思います．

II●2 DNA実験をする

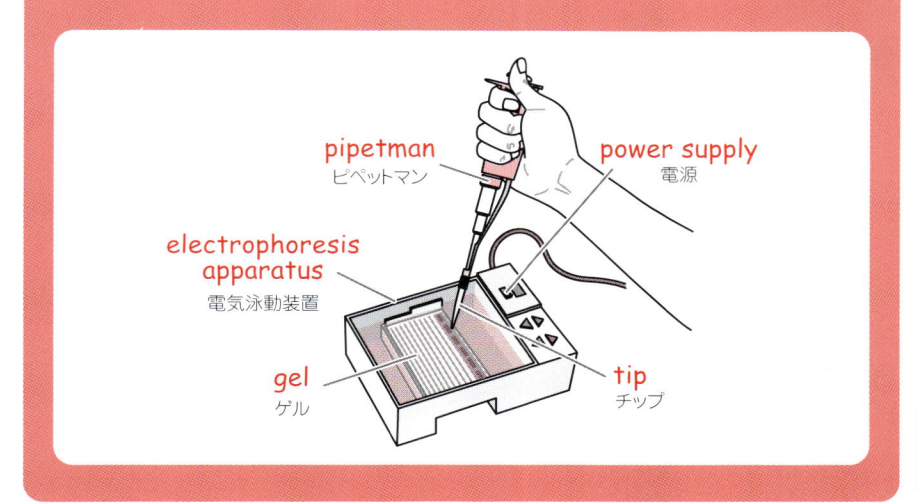

5 制限酵素処理と電気泳動

- pipetman ピペットマン
- power supply 電源
- electrophoresis apparatus 電気泳動装置
- gel ゲル
- tip チップ

1. 制限酵素処理/電気泳動
…*Restriction enzyme digestion/electrophoresis*

● Is it possible to perform a restrictive enzymatic reaction using more than 2 enzymes at the same time?

● 2種類以上の制限酵素で同時に反応させることは可能ですか？

● It is possible if the buffer conditions and reaction temperature of the restriction enzymes are the same.

● バッファーと反応温度が同じであれば可能です．

● How do you check the sample after digestion?

● 制限酵素処理後のサンプルをどのようにしてチェックしますか？

🐟 You check it by electrophoresis on agarose gel.

🦐 Why do you need to wear gloves when you make an agarose gel?

🐟 Because the gel contains a carcinogen (ethidium bromide).

🦐 **What's the percentage of** the agarose gel?

> 🦐/その他の表現法
> ◆ *What's the concentration of the agarose gel?*

🐟 This is a 1% agarose gel, and the percentage of the gel depends on the size of the target DNA that you are going to load.

🦐 What do you need to be careful about when you **apply samples**?

> 🦐/その他の表現法
> ◆ *What do you need to be careful about when you **put the samples in the wells**?*
> ◆ *When **loading a gel**, what precautions should you take?*

🐟 サンプルをアガロースゲルに流して確認します．

🦐 アガロースゲルを作製するときはなぜ手袋をつけるのでしょうか？

🐟 ゲルの中に発癌物質（エチジウムブロマイド）が含まれているからです．

🦐 そのアガロースゲルの濃度はどれくらいですか？

🐟 このゲルは1％ですが，この濃度は泳動するDNAのサイズによって決まります．

🦐 サンプルをウェルに入れるときに気をつけることは何ですか？

🐦 You should **take care not to** break the edge of a well, and you should avoid letting the samples leak out.

🐦 ウェルの端を壊さないように確実にサンプルをウェルに入れることです．

🐝/その他の表現法
◆ *You should <u>take care not to</u> puncture the edge of a well with the tip of the pipette and you should also be careful not to overfill it, as this will cause the sample to leak out.*

🐦 You can load samples into each well more easily if you **rest your elbows** on the bench.

🐦 ベンチの上に両肘をついてサンプルを入れると入れやすくなります．

🐝/その他の表現法
◆ *<u>Placing your elbows</u> on the bench top makes loading the samples easier.*
◆ *Loading the samples is easier if you <u>place your elbows</u> on the bench top.*
◆ *<u>Placing one's elbows</u> on the bench top facilitates easier gel loading.*
（非常に格式ばった言い方）

II・2 DNA実験をする

Column

実験台も背が高い！

　実験台の高さにもお国柄があります．すべてがそうかは知りませんが，私が留学していたアメリカのラボでは実験台の高さが90cmでした．日本の実験台の高さの標準は75〜80cmですので，最初大分高く感じました．しかし椅子にどっかり座って実験をする人は少なく，大部分の人が立ったままか，ハイチェアーにちょっと腰掛けるような感じで実験をしていました．ちなみに現在の私が所属する日本のラボの実験台の高さは90cmです．

Section II : Let's speak! 〜実験中の英会話

II●2 DNA実験をする

6 サブクローニング

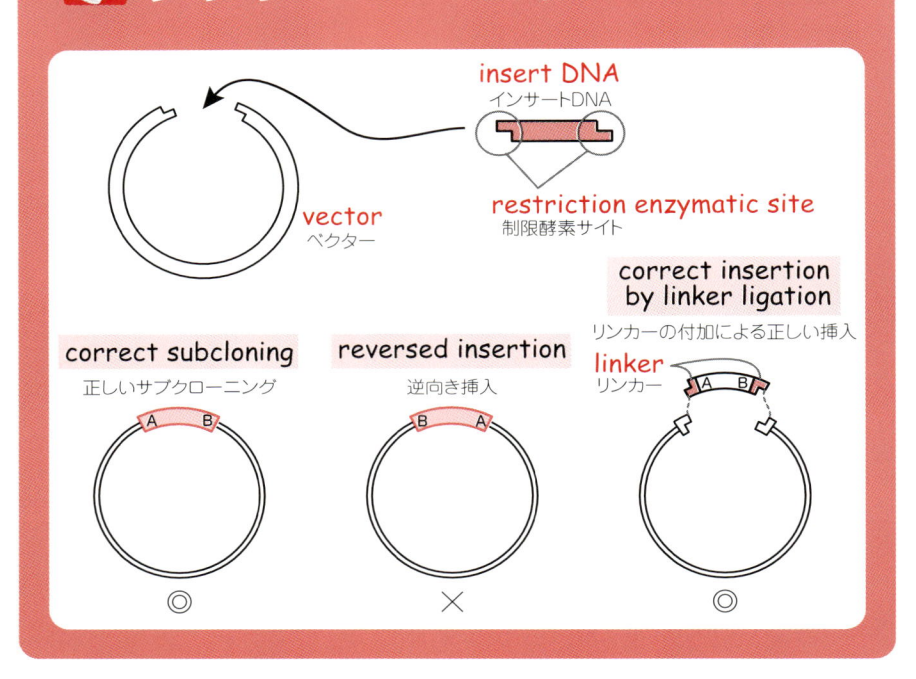

insert DNA
インサートDNA

vector
ベクター

restriction enzymatic site
制限酵素サイト

correct subcloning
正しいサブクローニング

reversed insertion
逆向き挿入

correct insertion by linker ligation
リンカーの付加による正しい挿入

linker
リンカー

1. インサートDNAの準備
…Preparing the insert DNA

● How do you extract insert DNA?

● You extract insert DNA **after cutting out a band** from the agarose gel.

その他の表現法
◆ *You extract insert DNA <u>after you remove a band</u> from the agarose gel.*

● インサートDNAはどのようにして抽出しますか？

● アガロースゲルからバンドを切り出してから抽出します。

52　バイオ実験 英語でトライ！

- How do you cut out the band from the agarose gel?
- バンドの切り出しはどのようにして行いますか？

- You put on safety glasses and do it on a transilluminator in a dark room.
- 暗室で，防護眼鏡を装着し，トランスイルミネーター上で行います．

2. ベクター調整
…Preparation of vectors

- How do you prepare a vector?
- ベクターはどのようにして調整しますか？

- You **perform** a **restriction reaction**, electrophorese the sample, and cut out a band from the agarose gel.
- 制限酵素で処理後，電気泳動しアガロースゲルからバンドを切り出します．

One Point
◆ perform：実験操作等を行うときよく使います．
examine, do experimentなどはあまり使いません．

◆ restriction reaction：制限酵素処理
restriction enzyme：制限酵素

- **What method do you use** to amplify **a small amount of commercially available vector?**
- 購入した少量のベクターをどのようにして増やしますか？

One Point

◆ *what method do you use~*.
実験の手法をたずねるときに使います．

その他の表現法

◆ *How do you amplify <u>a small amount of commercially available vector?</u>*
◆ *How do you amplify <u>a small amount of vector purchased from a supplier?</u>*
◆ *How do you amplify <u>a small amount of vector procured from a supplier?</u>*

🐟 You perform transformation, **amplification in liquid media**, and purification of DNA.

🐟 トランスフォーム後，液体培地で増やし，DNAを精製します．

その他の表現法

◆ *You perform transformation, <u>culture in liquid media</u>, and purification of DNA.*

3. ライゲーション
…Ligation

🐚 Under what conditions do you perform ligation?

🐚 ライゲーションはどのような条件で行いますか？

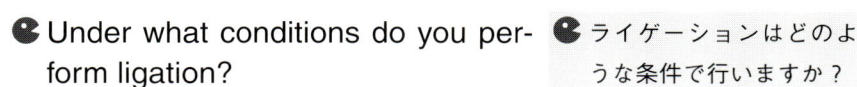

🐟 (You perform ligation) at 16℃ **overnight**.

🐟 一般に16℃でオーバーナイトで行います．

54 バイオ実験 英語でトライ！

One Point
◆ 一昼夜かけてインキュベーションするとき等に使います．O/Nと略して表示されていることもあります．

● How do you handle the sample after ligation?
● ライゲーション後のサンプルはどのように扱いますか？

● You **perform transformation**, amplification in liquid media, and purification of DNA.
● トランスフォーム後，液体培地で増やし，DNAを精製します．

その他の表現法
◆ *First you transfer it, then you amplify it in liquid media, and then you purify the DNA.*

4. インサートDNAのチェック
…Checking the size of the inserted DNA

● What method is used to **check the size of** the insert DNA?
● インサートDNAのサイズのチェックはどのようにして行いますか？

その他の表現法
◆ *How do you check the size of the insert DNA?*

● You **digest** the purified DNA after ligation, and electrophorese the digested DNA.
● ライゲーション後，精製したDNAを制限酵素処理し電気泳動します．

One Point
◆digest : perform restriction reaction（制限酵素処理）は単純にこのように表現されることもあります．

その他の表現法
◆You digest the purified DNA after ligation, and electrophorese the cut DNA by restriction enzyme.

● What would you think if the size of the DNA were different from the predicted size?

● 電気泳動の結果，予想サイズと異なるときどんなことを考えますか？

● Depending on the situation, it's possible that the insert DNA was placed **in the opposite or "reverse" orientation**.

● 場合によって，インサートDNAが逆向きに挿入されている可能性を考えます．

その他の表現法
◆Depending on the situation, it's possible that the insert DNA was placed *the wrong way round*.

II●2 DNA実験をする

7 形質転換（トランスフォーメーション）

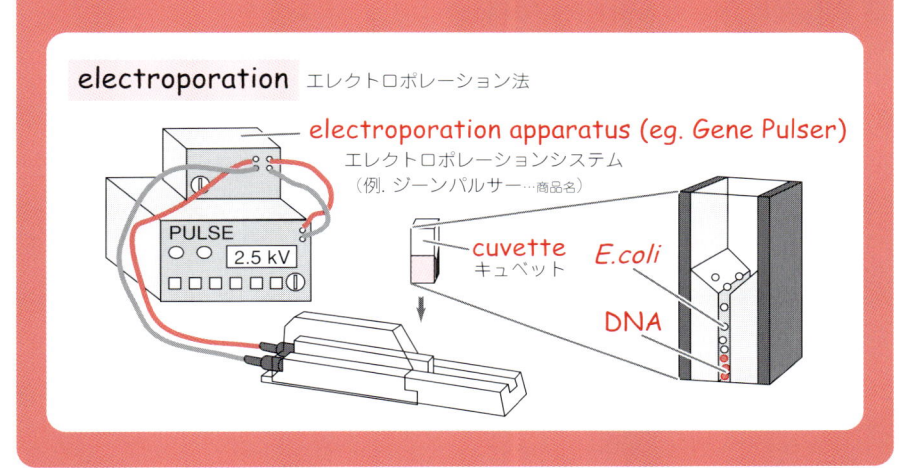

electroporation エレクトロポレーション法

electroporation apparatus (eg. Gene Pulser)
エレクトロポレーションシステム
（例．ジーンパルサー…商品名）

PULSE 2.5 kV

cuvette キュベット

E.coli

DNA

1．塩化カルシウム法
…Calcium chloride

🐟 **How do you** mix DNA solution and Calcium Chloride solution?

　🐟 DNA溶液と塩化カルシウム溶液を混和するときどのようなことに気をつけますか？

その他の表現法
　◆ <u>What do you need to be careful about</u> when you mix DNA solution and Calcium Chloride solution?

🐟 You mix them gently on a vortex, **trying to avoid making large particles**.

　🐟 緩やかにVortexにかけて，粒子が大きくならないようにします．

その他の表現法
　◆ <u>You mix them gently on a vortex taking care not to allow large parti-</u>

Section II：Let's speak! 〜実験中の英会話

cles to form.

🍫 What is the advantage of using the calcium chloride method?

🍭 **It can be performed easily** without expensive reagents or equipment.

🍫/その他の表現法
◆_It's easy to do_ and doesn't need expensive reagents or equipment.

🍫 塩化カルシウム法の特徴は何ですか？

🍭 高価な試薬や機械を必要とせず，簡便に行えることです．

2. エレクトロポレーション法
…Electroporation

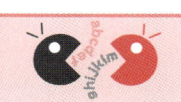

🍫 How do you handle the cuvettes before electroporation?

🍭 It's best to keep them on ice (at around 4℃).

🍫 What is the advantage of using the electroporation method?

🍭 It can introduce a large amount of DNA into the cells in one simple operation.

🍫 施行前のキュベットをどのように扱いますか？

🍭 氷上（4℃）で保存します．

🍫 エレクトロポレーション法の利点は何ですか？

🍭 簡単な操作で一度に大量のDNAを導入することができることです．

II・2 DNA実験をする

8 発現解析を行う

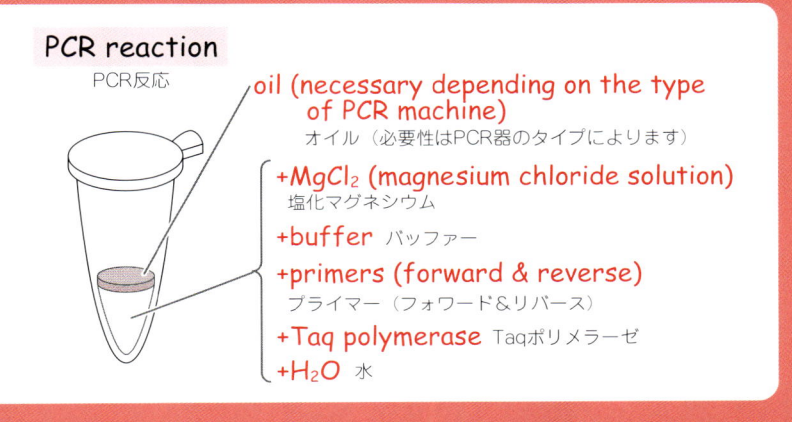

PCR reaction
PCR反応

- oil (necessary depending on the type of PCR machine)
  オイル（必要性はPCR器のタイプによります）
- +MgCl₂ (magnesium chloride solution)
  塩化マグネシウム
- +buffer バッファー
- +primers (forward & reverse)
  プライマー（フォワード＆リバース）
- +Taq polymerase Taqポリメラーゼ
- +H₂O 水

1. PCR法-① プライマーのデザイン
…Designing the primer

🍴 **How do you design** the primer for DNA sequencing?
🍴 配列はどのように決定しますか？

その他の表現法
◆ *What do you need to think about* when you *design* the primer?

🍴 You design the primer taking into consideration the percentage of GC content.
🍴 GCの含有率を考慮して決めます．

🍴 **Is there** any computer software **available**?
🍴 利用できるコンピュータソフトがありますか？

その他の表現法
◆ *Is there any computer software you can use*?

🐟 There is software for designing a primer.

🐟 プライマーデザインのためのコンピュータソフトがあります．

---

2. PCR法−②PCR装置の設定
…Setting the PCR machine/ Setting the conditions for the PCR machine

🐟 What **temperature settings need to be made**?

🐟 必要な温度設定は何ですか？

その他の表現法
◆ *What are the required temperature settings?*

🐟 It's necessary to set the temperature for denaturing, annealing, and extension.

🐟 Denaturing, annealing そしてextensionの温度です．

🐟 How do you store samples after a PCR reaction?

🐟 PCR反応終了後のサンプルは，どのように保存しますか？

🐟 You store samples at 4℃ until electrophoresis.

🐟 電気泳動するまで4℃で保存します．

3. ノーザンハイブリダイゼーション
…Northern hybridization

🔴 Why do you use denaturing agarose gel here? It is not used for DNA.

🔴 DNAの場合と異なり，変性アガロースゲルを用いるのはどうしてですか？

🔴 Because it **reflects** the molecular weight of single-stranded RNA **more accurately**.

🔴 この方が１本鎖RNAの分子量を正確に反映するからです．

> 📝 その他の表現法
> ◆ Because it *shows* the molecular weight of single-stranded RNA *more precisely*.

🔴 **What steps are required** to produce single-stranded DNA from a radioisotope-labeled DNA probe before hybridization?

🔴 ラジオアイソトープで標識したDNAプローブをハイブリダイゼーションの前に１本鎖にするため，どのようなステップを行いますか？

> 📝 その他の表現法
> ◆ *What actions do you need to take to produce single-stranded DNA from a radioisotope-labeled DNA probe before hybridization?*

🔴 You boil the DNA probe at 100℃ and then cool it down quickly in ice for 5 minutes.

🔴 DNAプローブを100℃で煮沸後，氷中で５分間急冷します．

4. RNAase プロテクションアッセイ
…RNA protection assay

● **How does** RNAase protection assay **compare to** Northern hybridization in terms of detection sensitivity?

　◆ その他の表現法
　◆*How does RNAase protection assay measure up to Northern hybridization in terms of detection sensitivity?*

● The RNAase protection assay is better.

● Why is it better?

● Because an RNA probe is used in an RNAase protection assay.

● RNAaseプロテクションアッセイの検出感度はノーザンハイブリダイゼーションと比べてどうですか？

● RNAaseプロテクションアッセイの方がよいです．

● その理由は何ですか？

● RNAaseプロテクションアッセイではRNAプローブを使うためです．

II・2 DNA実験をする

9 塩基配列の決定（DNAシークエンス）

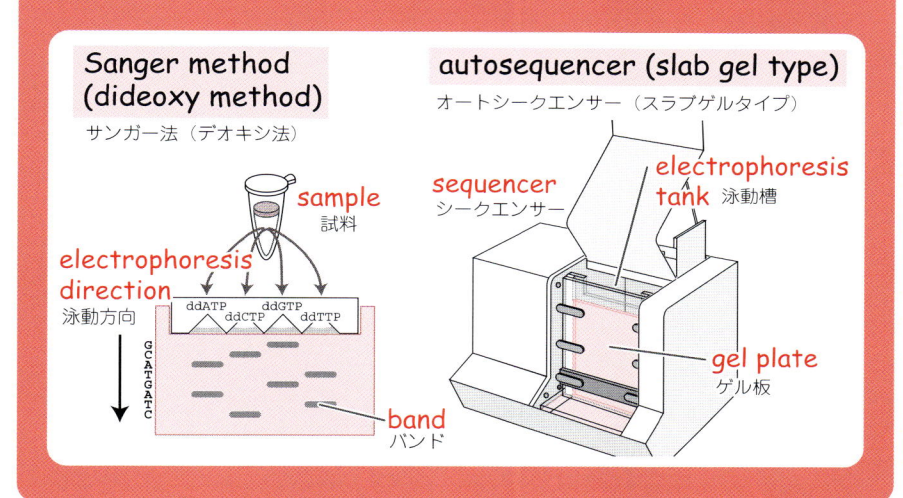

Sanger method (dideoxy method)
サンガー法（デオキシ法）

- sample 試料
- electrophoresis direction 泳動方向
- ddATP, ddCTP, ddGTP, ddTTP
- band バンド

autosequencer (slab gel type)
オートシークエンサー（スラブゲルタイプ）

- sequencer シークエンサー
- electrophoresis tank 泳動槽
- gel plate ゲル板

1. 電気泳動
…Electrophoresis

● What do you have to be careful about when you prepare a gel for electrophoresis?

● 電気泳動のためのゲルを作製するとき気をつけることは何ですか？

● Since there is a large amount of gel, (OR Since the gel is **thick**,) you should be careful **not to let air bubbles get in**.

● ゲルが大きいため、気泡が入らないよう気をつけます．

その他の表現法
◆ Since there is a large amount of gel, (OR Since the gel is *thick*,) you should *avoid letting air bubbles in*.

Section II：Let's speak! ～実験中の英会話 63

● What should you do about the gel after you have finished electrophoresis?

● 電気泳動終了後のゲルはどうしますか？

● You should dry it with gel drier and **expose it to X-ray film.**

● ゲルドライヤーで乾燥させた後，X線フィルムに感光させます．

その他の表現法
◆ *You should dry it with gel drier and X-ray it.*

---

2. オートシークエンサーを用いた解析
…Analysis of the DNA sequence using an autosequencer

● What's the advantage of using an autosequencer?

● オートシークエンサーの利点は何ですか？

● The advantage is that we can **analyze** several samples **in a short time**.

● 数種類のサンプルを短時間に解析できることです．

その他の表現法
◆ *The advantage is that we can study several samples in a short period of time.*
◆ *The advantage is that we can examine several samples in a short space of time.*

● How do you analyze the data obtained by autosequencing?

● データ解析はどのようにして行いますか？

🌀 You analyze them using computer software.

その他の表現法
◆ *You use computer software.*

🌀 コンピュータソフトを使用して解析します．

Column 勇気を出して

　多くのラボの人とコミュニケーションを取っていく上で名前の呼び方は大事です．特にわれわれ日本人は，案外，直接名前を言わないことも多いかと思います．ご存じの通りアメリカではファーストネームで呼ぶのが普通で，Bossであれ，目上の人であれ，普通は呼び捨てです．初めてラボを訪れたとき，「Proffesor（Last name）ー！」と後ろから呼び掛けたのですが振り向いてもらえませんでした．ただ単に忙しかっただけだったのかもしれませんが，2回目に勇気を出して，ファーストネームの呼び捨てで呼び掛けたところ，「調子はどうだい？」と笑顔で聞き返してくれました．日本の教授の前ではとても言えないことですが，アメリカではこれが普通です．

Column 長蛇のレジ！その理由は…

　留学中は，手の空いたラボのメンバーといつも病院のカフェテリアに行ってランチを楽しみました．そこはビュッフェタイプで，好きなものを自分で選んで最後にレジで精算する方式でした．2人いるレジの担当の人のうち，男性の方はいつもすいているので，ある日そちらで精算した後，みんなと話をしていてその理由がわかりました．全く同じものを選んだのに値段が相当違っていたのです（といっても，全額でせいぜい＄5くらいまでのランチなのですが）．次回から，長蛇の列の方に並んだのは言うまでもありません．研究で留学されている方はおおむね，厳しい経済状態で日々過ごされています．少しでも長く滞在するために，こんな情報でも実験のために必要となってくるのです．

II・2 DNA実験をする

10 データ解析とホモロジー検索

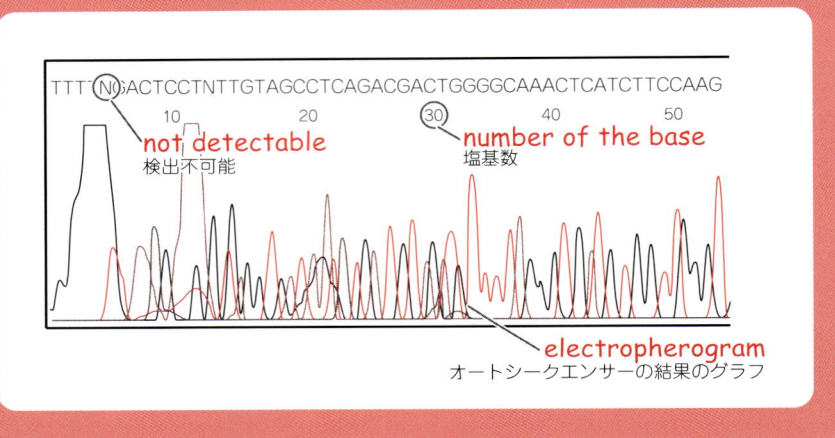

```
TTTTN(N)ACTCCTNTTGTAGCCTCAGACGACTGGGGCAAACTCATCTTCCAAG
    10        20       (30)      40        50
```
not detectable
検出不可能

number of the base
塩基数

electropherogram
オートシークエンサーの結果のグラフ

1. エレクトロフェログラム …Electropherogram

🍂 What is an electropherogram?

🍂 エレクトロフェログラムとは何ですか？

🟠 It gives data on base sequences obtained by an autosequencer.

🟠 オートシークエンサーにより得られた塩基配列のデータです．

🍂 **What are the disadvantages** of sequence data obtained by electrophoresis?

🍂 電気泳動により得られた配列データの欠点は何ですか？

🐾 その他の表現法
◆ <u>What is the drawback</u> of sequence data obtained by electrophoresis?

🐢 It takes a long time to analyze the data and occasionally detects non-specific bands.

🐢 データ解析に時間がかかることと，場合によって非特異的バンドが検出されることです．

2. データベース
…Database

🐢 How can you access information on **known** gene sequences?

🐢 既知の遺伝子配列の情報はどのようにして知ることができますか？

その他の表現法
◆ How can you access information on *recognized* gene sequences?

🐢 You can **get data on-line** from a gene information web site such as GeneBank.

🐢 GeneBankなどの遺伝子情報サイトからオンラインでデータを取りだせます．

その他の表現法
◆ You can *go on-line and obtain data* from a web site such as GeneBank.

🐢 How do you carry out a homology search?

🐢 ホモジー検索はどのようにして行いますか？

🐢 You use gene analysis software.

🐢 遺伝子解析ソフトを用いて行います．

Chapter 3
タンパク質実験をする

▶ 実験の流れに添って各ステップで役立つ英語表現を紹介しています．

1 タンパク質の精製 69
　　1. タンパク質の抽出／2. タンパク質の分離精製（透析）

2 精製タンパク質の分析 71
　　1. タンパク質の定量／2. タンパク質の分子量測定

3 タンパク質の活性測定 73
　　1. DNA結合活性測定／2. DNA結合部位決定法

4 タンパク質の機能調節の解析 75
　　1. ウエスタンブロッティング／2. 免疫沈降法／
　　3. Two-ハイブリッドシステム

5 タンパク質の分布の解析 78
　　1. *in situ*での検出／2. *in vitro*での検出

II●3 タンパク質実験をする

❶ タンパク質の精製

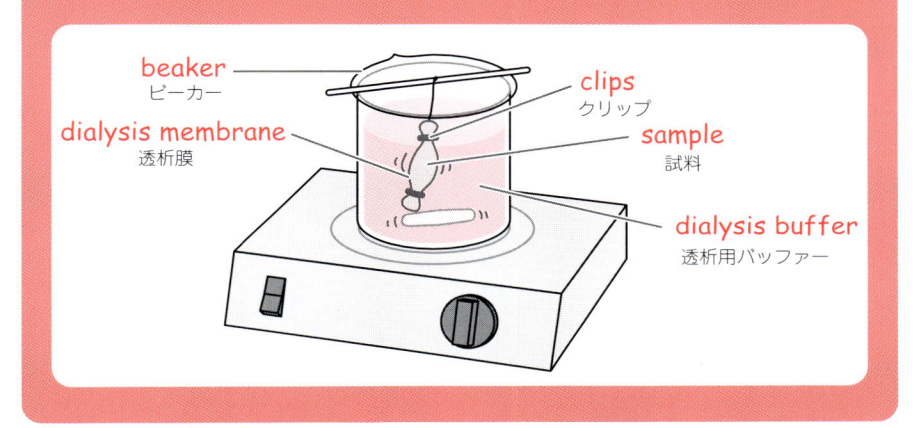

- beaker ピーカー
- dialysis membrane 透析膜
- clips クリップ
- sample 試料
- dialysis buffer 透析用バッファー

1. タンパク質の抽出
…Extracting protein

● How do you extract nuclear protein?

● 核タンパク質の抽出はどのように行いますか？

● Generally speaking, you add extract solution, and perform **freeze-thaw cycles**.

● 一般に抽出液を添加後，凍結，融解を繰り返して行います．

● What method is used for extracting a specific protein?

　◆その他の表現法
　◆ How do you extract a specific protein?

● 特定のタンパク質を抽出するためにどのような方法を用いますか？

● One option is to use chromatography.

● クロマトグラフィーを用います．

2. タンパク質の分離精製（透析）
…Protein purification (dialysis)

🍮 **What's the purpose of** dialysis?

🐛 その他の表現法
◆ *What's the point of* performing dialysis?

🍮 The purpose is to remove **contamination**.

🐛 その他の表現法
◆ *The purpose is to remove impurities.*

🍮 透析はどのような目的で行われますか？

🍮 不純物を除去する目的で行われます．

🍮 **What kind of equipment** is used for dialysis?

🐛 その他の表現法
◆ *What type of apparatus* is used for dialysis?

🍮 The equipment comprises dialysis membranes, clips, beakers, dialysis buffer, etc.

🍮 どのような器具を用いますか？

🍮 透析膜，クリップ，ビーカー，透析用バッファーなどです．

II●3 タンパク質実験をする

2 精製タンパク質の分析

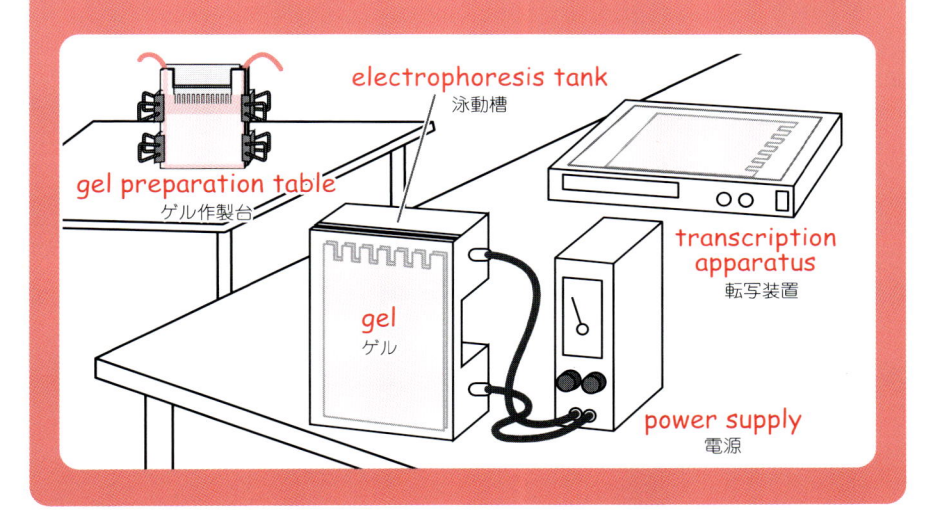

electrophoresis tank 泳動槽
gel preparation table ゲル作製台
gel ゲル
transcription apparatus 転写装置
power supply 電源

1. タンパク質の定量
…Protein assay

🎃 **What are the standard assays** for protein?

　🎀 その他の表現法
　◆ *What kind of standard protein assays are available?*

🌀 The standard ones include the Lowry, BCA and Bradford assays.

🎃 タンパク質の定量でスタンダードな方法は何ですか？

🌀 Lowry法やBCA法やBradford法があります．

🎃 What is the advantage of using the SDS-PAGE method?

🎃 SDS-PAGE法の特徴は何ですか？

Section II：Let's speak! 〜実験中の英会話

🐦 It is **less affected** by interfering substances.

🐦 妨害物質の影響を受けにくいことです．

🦐 その他の表現法
◆ Interfering substances *have less of an effect* with this method.

2. タンパク質の分子量測定
…Determination of protein molecular weight

🐤 What methods exist to determine protein molecular weight?

🐤 タンパク質の分子量測定にどのような方法がありますか？

🐦 There are the SDS-PAGE and gel filtration methods.

🐦 SDS-PAGE法やゲル濾過法があります．

🐤 Are **the values** you obtain from prestain markers used in the SDS-PAGE method **accurate**?

🐤 SDS-PAGE法に用いるプレステインマーカーの値は正確ですか？

🦐 その他の表現法
◆ Are *the figures* you obtain from prestain markers used in the SDS-PAGE method *exact*?

🐦 No, they are not very accurate.

🐦 あまり正確ではありません．

II●3 タンパク質実験をする

3 タンパク質の活性測定

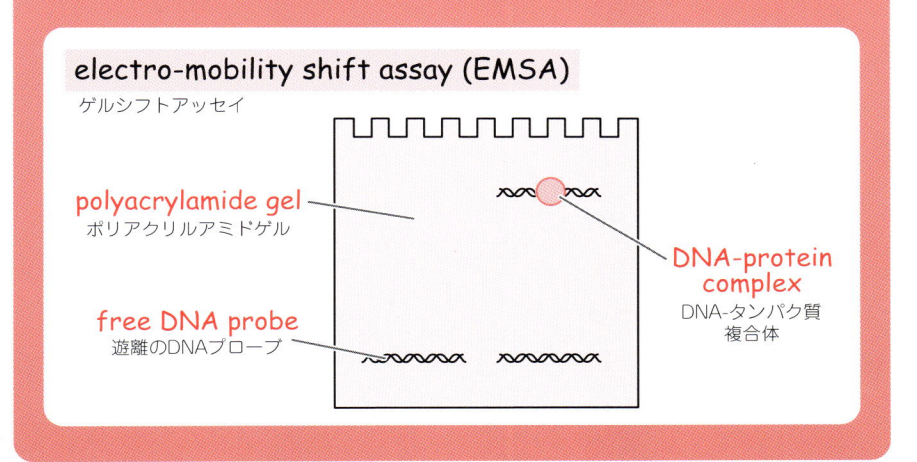

electro-mobility shift assay (EMSA)
ゲルシフトアッセイ

- polyacrylamide gel / ポリアクリルアミドゲル
- DNA-protein complex / DNA-タンパク質複合体
- free DNA probe / 遊離のDNAプローブ

1. DNA結合活性測定
…Determination of DNA binding activity

● What methods can be used to determine DNA binding activity?

● DNA結合活性測定にはどのような方法がありますか？

● You can use the electro-mobility shift assay (EMSA).

● Gel shift法があります．

● **How does the electro-mobility shift assay determine** DNA binding activity?

● Gel shift法ではどのようにしてDNA結合活性を測定しますか？

🔹 その他の表現法
◆ *How does the electro-mobility shift assay measure* DNA binding activity?

🐠 It **analyzes the interaction** of DNA with a transcription factor.

🐚 その他の表現法
◆ It *examines how* transcription factors *interact with* DNA.

🐠 DNAと転写調節因子との相互作用を解析することにより測定します．

2. DNA結合部位決定法
…Determination of DNA binding sites

🐡 What methods are available to determine DNA binding sites?

🐡 DNA結合部位決定法にはどのような方法がありますか？

🐠 Foot print analysis is one such method.

🐠 Foot print法があります．

🐡 **What kind of enzyme is used** in foot print analysis?

🐚 その他の表現法
◆ *What enzymes are used* in foot print analysis?

🐡 Foot print法で用いる酵素は何ですか？

🐠 You use DNAase Ⅰ.

🐠 DNAase Ⅰを用います．

II・3 タンパク質実験をする

❹ タンパク質の機能調節の解析

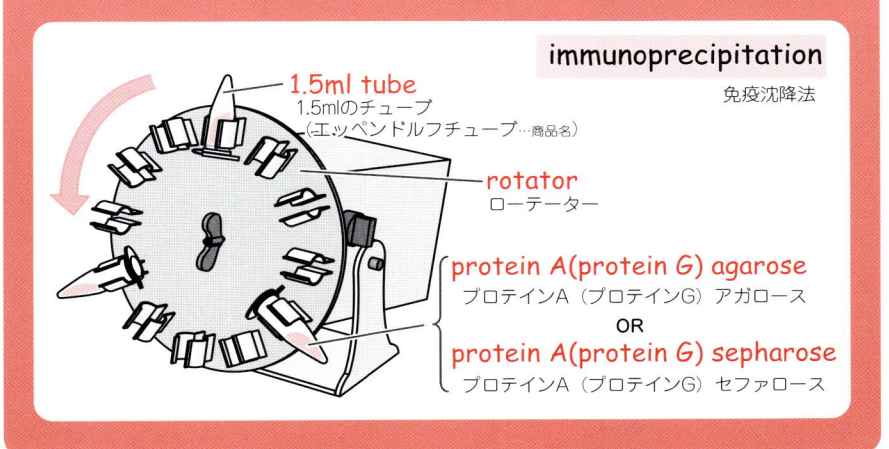

immunoprecipitation
免疫沈降法

- 1.5ml tube
  1.5mlのチューブ
  (エッペンドルフチューブ…商品名)
- rotator
  ローテーター
- protein A(protein G) agarose
  プロテインA（プロテインG）アガロース
  OR
- protein A(protein G) sepharose
  プロテインA（プロテインG）セファロース

1. ウエスタンブロッティング
…Western blotting

🍽 **What kinds of** membranes are used for Western blotting?

その他の表現法
◆ *What types of* membranes are there for Western blotting?

ウエスタンブロッティングで用いるメンブレンにはどのようなものがありますか？

🍽 These include nitrocellulose membrane, PVDF membrane and so forth.

ニトロセルロース膜やPVDF膜等があります．

🍽 What **methods** are there **to label the antibodies** used in Western blotting?

ウエスタンブロッティングで用いる抗体の標識法にはどのようなものがあ

Section II : Let's speak! 〜実験中の英会話

🐛 その他の表現法
◆ *How do you label the antibodies* in Western blotting?
◆ *What means* are employed *to label the antibodies* used in Western blotting?（非常に格式的！）

🐦 They are labeled with radioisotopes or enzymes.

🐛 その他の表現法
◆ *You label them with radioisotopes or enzymes.*

🐦 放射線同位元素標識や，酵素標識があります．

2. 免疫沈降法
…*Immunoprecipitation*

🎧 What is the purpose of immunoprecipitation?

🎧 免疫沈降法はどのような目的で使われますか？

🐦 It is used to **identify proteins** that interact with the antibody.

🐦 相互作用するタンパク質の同定に用いられます．

🐛 その他の表現法
◆ *It is used to ascertain which proteins interact with the antibody.*

🎧 What is an immunocomplex?

🎧 免疫複合体とは何ですか？

🐦 It is a complex of extracted protein

🐦 抽出したタンパク質と目的

and the antibody that recognizes the target protein.

のタンパク質を認識する抗体の複合体のことです．

3. Two-ハイブリッドシステム
…Two-hybrid system

🌑 What is the advantage of this system?

🌑 このシステムの特徴は何ですか？

🌑 It can detect the target proteins with a high level of sensitivity from a library of proteins that interact with them.

🌑 目的のタンパク質を相互作用のあるタンパク質のライブラリーから高感度に検出できる点です．

Column: Surprising Party（Birthday）

　アメリカ人のバースデーに対する思いはかなり強いものがあります．アメリカのラボのテクニシャンは友人がつくってきた紙製の冠？を朝からかぶってはしゃいでいました．ケーキはお世辞にもおいしいとは言えませんがBirth dayをみんなで祝うその気持ちが大事なのでしょう．もう1つ，partyの直前まで本人に知らせずにおくこと．筆者は『xx時にmeeting roomへ集まって』と言われ，その目的を何も知らずに誕生日の人に事務的に伝えてしまいました．後でみんながあきれた顔をしてmeeting roomで私の方を見ていたのは言うまでもありません．アメリカでは実験の手順を覚えておくことも大事ですが，いつもカレンダーをチェックして他人のBirth dayも覚えておくことが大切です．

II●3 タンパク質実験をする

5 タンパク質の分布の解析

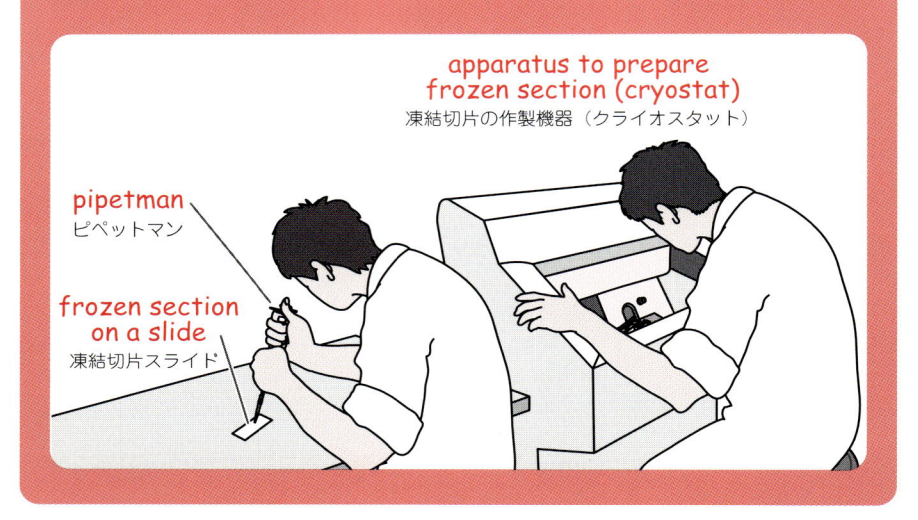

apparatus to prepare frozen section (cryostat)
凍結切片の作製機器（クライオスタット）

pipetman
ピペットマン

frozen section on a slide
凍結切片スライド

1. *in situ* での検出
…in situ detection

● What method can be used to fix the target protein?

● 目的のタンパク質の固定法にはどのようなものがありますか？

● You can fix it with acetone or alcohol.

● 凍結，アセトンやアルコールによる固定法があります．

● If signals cannot be detected by using an antibody, **what should you try to do**?

● 抗体でシグナルが検出されない場合，何を試してみますか？

> **その他の表現法**
> ◆*If signals cannot be detected by using an antibody, <u>what else can you do</u>?*

🔴 You should try to **activate its antigenicity**, by treating it in a microwave oven, for example.

> **その他の表現法**
> ◆*You should try to <u>enhance its antigenicity</u>, by treating it in a microwave oven, for example.*

🔴 電子レンジで処理するなどして，抗原性を賦活化してみます．

2. *in vitro* での検出
…*in vitro* detection

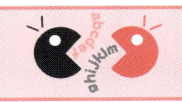

🟤 What **problems might arise** when you try to detect protein *in vitro*?

🟤 *in vitro* での検出でタンパク質を検出する際，問題点は何ですか？

> **その他の表現法**
> ◆*What <u>points do you have to bear in mind</u> when you try to detect protein in vitro?*

🔴 When the protein is below the detectable range, you need to concentrate it.

🔴 検出限界以下の場合，タンパク質を濃縮する必要があります．

🟤 How do you concentrate the protein using salts?

🟤 塩による濃縮を行う場合，どのようにしますか？

🐤 After the protein is eluted with ammonium sulfate, it is dialyzed and then analyzed.

🐤 硫酸アンモニアを用いて析出させた後，透析して解析に用います．

Column 切り換えも大切？

　アメリカの朝は早いです．明け方から車の音が急に騒がしくなります．アメリカが日本以上に車社会なので余計にそう感じるのかもしれません．しかし，帰りも早いです．金曜日は早々に仕事を切り上げてさっさと帰る人も結構います．私の留学先では，秘書さん（といっても数人いる内の第１秘書）はいつも金曜日からお休みでした．われわれも彼等のペースにあわせる必要はないと思いますが，場所によっては治安の悪い地域にあるラボもありますから，あまり深夜まで実験しない方が安全かもしれません．しかし，中にはリズムが全く反対で，われわれが帰る頃ラボにやってきて，翌朝まで自由自在に実験をするタイプの人もいました．実験器具の予約の必要がなく，自分のペースで実験できる長所がありますが，体調を崩す可能性もあり一般にはお薦めできません．

Column 開かずの扉

　アメリカのラボの扉は頑丈にできていますが，いったん閉まると，鍵を使ってなかなかうまくあけられないことがあります．平日は誰かがいるので問題になりませんが，休みの日にラボへ行くと，鍵がかかっていて，なかなか開けられず右往左往していて，警備員に不審者と間違われそうになることがあります．解錠を一度試しておくことをお薦めします．あと，共通の機器が置いてある部屋の鍵が休日にはかかっていて入れないことがありました．休日に実験予定が入っている場合は気をつけましょう．

MEMO

Chapter 4
細胞培養をする

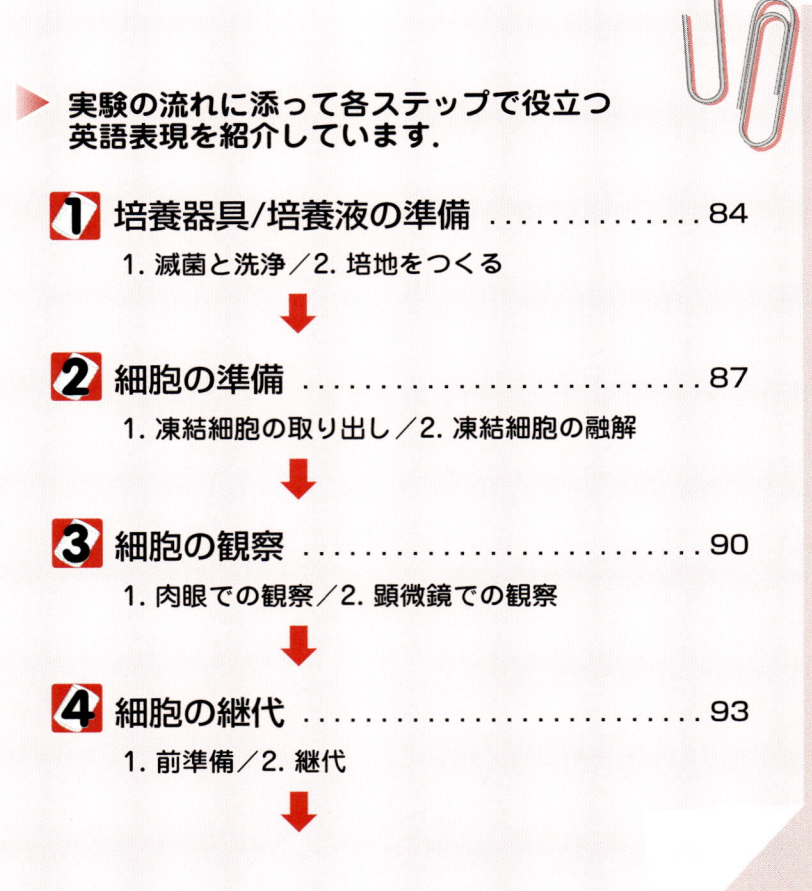

▶ 実験の流れに添って各ステップで役立つ英語表現を紹介しています．

1 培養器具/培養液の準備 84
　1. 滅菌と洗浄／2. 培地をつくる

2 細胞の準備 87
　1. 凍結細胞の取り出し／2. 凍結細胞の融解

3 細胞の観察 90
　1. 肉眼での観察／2. 顕微鏡での観察

4 細胞の継代 93
　1. 前準備／2. 継代

5 細胞数の測定 96
1. 細胞浮遊液の調整／2. 血球計算板での計測

6 細胞のクローニング 99
1. コロニーの選定／2. 細胞の回収

7 細胞の保存 102
1. 細胞懸濁液／2. 凍結

8 培養実験を終えて 105
1. 培養器具の処理／2. 培養液の処理

II●4 細胞培養をする

❶ 培養器具/培養液の準備

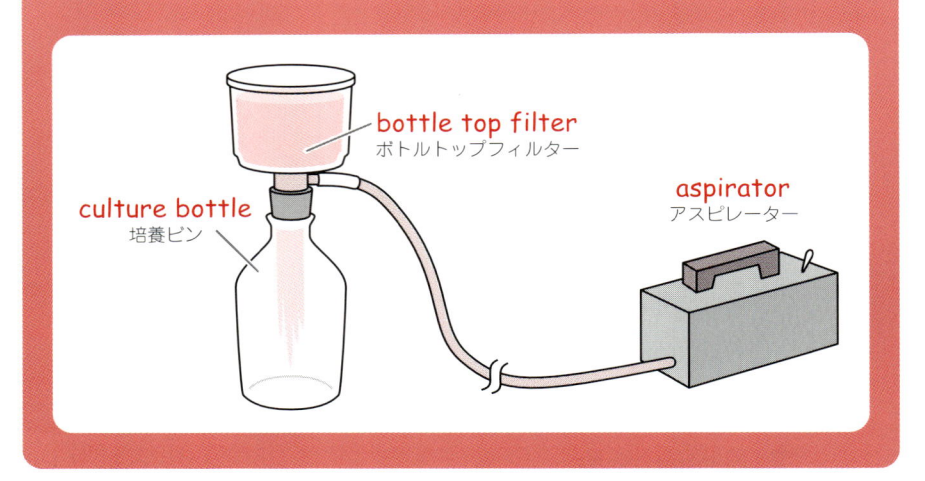

- bottle top filter
  ボトルトップフィルター
- culture bottle
  培養ビン
- aspirator
  アスピレーター

1. 滅菌と洗浄
…Sterilization and washing

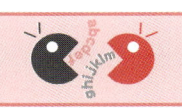

🍫 How should you sterilize cell culture medium **that has been prepared from powder**?

> 🐛 **その他の表現法**
> ◆ How should you sterilize cell culture medium <u>*made from powder*</u>?

🍫 粉末から作製した細胞培養培地はどのように滅菌しますか？

🍓 It should be filter sterilized.

🍓 フィルターを通して濾過滅菌します．

🍫 **Can you use** a reagent bottle that has contained liquid medium **again**?

🍫 液体培地が入っていた試薬ビンは再利用できますか？

84　バイオ実験 英語でトライ！

> 📣 その他の表現法
> ◆ **Can you recycle** a reagent bottle that has contained liquid medium?

🐟 Yes, you can, if you wash, dry, and autoclave it.

🐡 洗浄後、乾燥させ、オートクレーブをかけて再利用できます.

🐡 What do you do when you put a bottle containing medium in a clean bench?

🐡 培地の入った試薬ビンをクリーンベンチ内に入れるとき、どのようにしますか？

🐟 You first wipe the surface of the bottle with cotton soaked in alcohol.

🐟 アルコール綿で、表面を拭って持ち込みます.

2. 培地をつくる
…Preparation of culture media

🐡 **At what concentrations** should we add fetal bovine serum to the medium?

🐡 培地に加える牛胎仔血清の濃度はどれくらいですか？

> 📣 その他の表現法
> ◆ **What percentage** of fetal bovine serum should be added to the medium?

🐟 About 5% to 20%, depending on

🐟 培養細胞によって異なり

the cells to be cultured. | ますがだいたい 5 ％から 20％の間です．

🌑 Why is the medium red? | 🌑 培地が赤色をしているのはどうしてですか？

🌰 Because it contains phenol red. | 🌰 フェノールレッドが含まれているからです．

🌑 What can you tell from changes in the color of the medium? | 🌑 培地の色の変化でどのようなことがわかりますか？

🌰 You can guess the change in pH levels in the medium. | 🌰 培地のpHの変化が推測できます．

Column：コンタミについて（日本とボストンの培養環境のちがい）

　筆者がボストンのラボを訪れて，培養室を見て驚いたのは，培養液の入った瓶の口や，パスツールピペットを滅菌するためのガスバーナーがどこも見当たらないことでした．医療廃棄用の段ボール性の箱に，70％エタノールのプラボトルが無造作に掛けてあるだけでした（あるときは地面に転がっていました）．これだとコンタミが心配だなと思っていましたが，不思議と一度もコンタミは起こりませんでした．一方，日本では，アメリカよりも数段気を使って培養をしてもときにコンタミに遭遇していたと思います．アメリカの気候（特にボストン？）は日本に比べ相当湿度が低いため，このことが影響しているのかもしれません．

II●4 細胞培養をする

2 細胞の準備

water bath (37℃)
37℃の温浴槽

liquid nitrogen tank
液体窒素タンク

frozen cell vials
冷凍細胞バイアル

1. 凍結細胞の取り出し
…Retrieval of frozen cells

● What do you have to be careful about when you take out frozen cells?

● 凍結細胞を取り出すとき注意することはどんなことですか？

● You should **thaw them quickly** at 37℃ and place them in a medium to dilute them.

● 素早く37℃で融解し，培地の中へ移して希釈することです．

その他の表現法
◆ You should <u>defrost them promptly</u> at 37℃ and place them in a medium to dilute them.
◆ You should <u>unfreeze them rapidly</u> at 37℃ and place them in a medium to dilute them.

🟤 How should you open a vial that contains frozen cells?

🔴 You should only open it after you have wiped the cap with alcohol.

🟤 Should you take out frozen cells **with bare hands**?

> 🍀 その他の表現法
> ◆ *Should you take out frozen cells <u>without wearing gloves</u>?*

🔴 No, you must wear gloves because they have been kept in an ultra-low temperature, liquid nitrogen freezer.

🔴 凍結細胞の入ったバイアルを開けるとき，どのように扱いますか？

🔴 ネジ（キャップ）の部分をアルコールで拭ってから開けます．

🔴 凍結細胞を取り出すときは，素手で行いますか？

🔴 いいえ，超低温槽（液体窒素）から取り出すので専用の手袋をします．

2. 凍結細胞の融解
　　…Thawing frozen cells

🟤 Is it possible to thaw several vials at the same time?

🔴 **It is safer** to thaw them **one by one**, even if it takes longer.

🔴 凍結細胞を度一度に数本融解することは可能ですか？

🔴 手間がかかっても，1本ずつ行ったほうが無難です．

その他の表現法

◆ *It is wiser* to thaw them *one at a time*, even if it takes longer.

● After you have transferred the frozen cells to the medium, should you seed the cells in a culture flask as they are?

● 凍結細胞を培地へ移したら、そのまま培養フラスコに播きますか？

● Occasionally, depending on the type of cells, you centrifuge them once and then change the medium.

● 細胞の種類によりますが、一度遠心して培地交換することもあります．

● Why do you dilute frozen cells with a medium when you thaw them?

● 凍結細胞の融解時に、培地で希釈するのはどうしてですか？

● Because the vials for frozen cells may contain cytotoxic substances such as DMSO.

● 凍結細胞のバイアルに、細胞毒性を持った物質（DMSOなど）が含まれているからです．

II•4 細胞培養をする

Ⅱ●4 細胞培養をする

❸ 細胞の観察

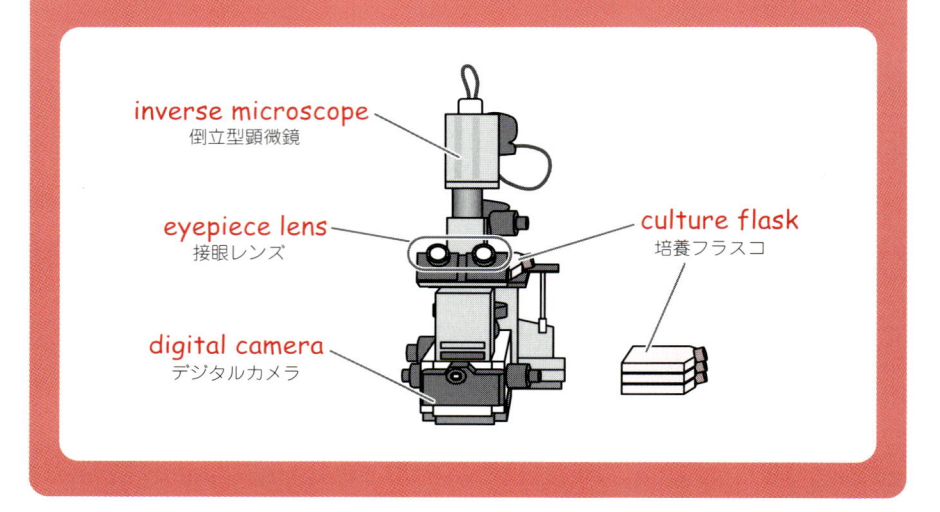

inverse microscope
倒立型顕微鏡

eyepiece lens
接眼レンズ

culture flask
培養フラスコ

digital camera
デジタルカメラ

1. 肉眼での観察
…Observation with the naked eye

● **What should you check first** when you observe cultured cells with the naked eye?

その他の表現法
◆ *What's the first thing to look for when you observe cultured cells with the naked eye?*

● You should check the color of the medium and look for any contamination.

● 培養細胞を肉眼で観察するとき，まずどのようなことをチェックしますか？

● 培地の色や，コンタミ (contamination) の有無をチェックします．

🐚 What would it mean if the medium had turned yellow?

🐚 培地の色が黄色になっていたら，どのようなことが考えられますか？

🐚 **One might surmise** that the cell density was very high and that you should subculture them.

🐚 細胞密度が非常に高く，継代が必要であることが推察されます．

🐚/その他の表現法
◆ *This could imply that the cell density was very high and that you should subculture them.*

🐚 How would you notice if the medium was contaminated?

🐚 コンタミを起こすとどうなりますか？

🐚 It would be turbid and the smell would be different.

🐚 培地が濁ったり，場合によっては培地のにおいが変わります．

---

2. 顕微鏡での観察
…*Microscopic observation*

🐚 What should you be careful about when you **observe** cultured cells **under a microscope**?

🐚 培養細胞を顕微鏡で観察するとき，どのようなことに注意しますか？

🐚/その他の表現法
◆ *What should you be careful about when you **take a look at** cultured cells **with a microscope**?*

Section Ⅱ：Let's speak! ～実験中の英会話

🔴 Normally, you should start from a low magnification.

🟢 通常は，まず低倍率から観察をはじめます．

🟢 How should you record the cells you observe under a microscope?

🟢 顕微鏡で観察した培養細胞の記録はどのようにして残しますか？

🔴 You either draw a sketch or take a photo with a camera attached to the microscope.

🔴 スケッチをするか，顕微鏡にカメラを取り付けて写真で残します．

🟢 How should you handle the culture flask when you observe the cells under the microscope?

🟢 顕微鏡で観察するとき培養フラスコをどのように扱いますか？

🔴 You should be careful not to tilt it or the cap will come into contact with the medium.

🔴 傾けたりして，培地がキャップに触れないように気をつけます．

Column アイソトープも一般の実験室で？！

日本の実験施設でアイソトープを使うためにはアイソトープ管理区域で種々の規則のもと使用しなければなりません．ただし，アメリカでは一般の実験室の一角でも使用が可能です．当然防護板等は必要ですが，日本に比べると規定がずっと緩い感じがしました．ただし，ときどき見なれない人が実験室にやってきてガイガーカウンターで抜き打ちの検査にきていました．

II•4 細胞培養をする

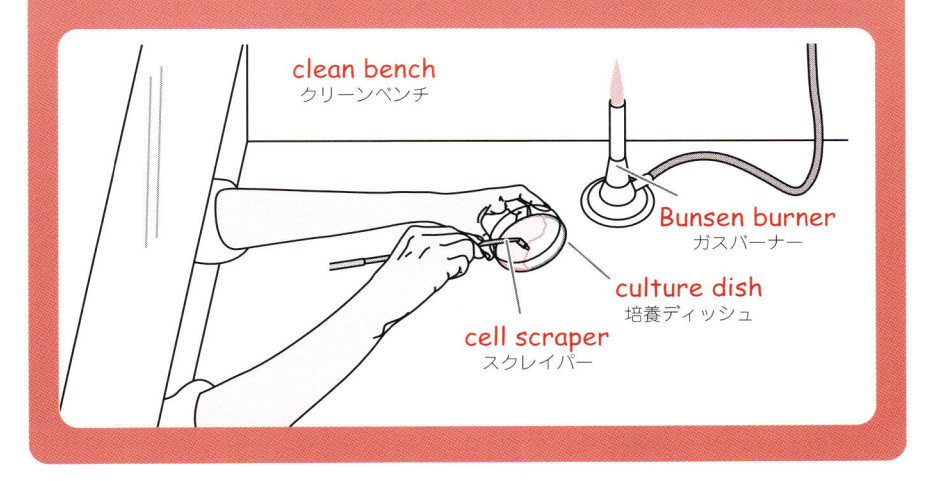

❹ 細胞の継代

- clean bench クリーンベンチ
- Bunsen burner ガスバーナー
- culture dish 培養ディッシュ
- cell scraper スクレイパー

1. 前準備
…Preparation

- **Tell me what you need** to subculture cells.

  ◆*What do you need* to subculture cells?

- You need PBS (phosphate buffer saline) and trypsin solution.

- **Can you use** PBS and trypsin solution kept at 4℃ **as is**?

― 細胞の継代に必要なものを挙げてください．

― PBS（phosphate buffer saline）とトリプシン液です．

― PBSとトリプシン液は4℃のままで使用しますか？

その他の表現法

◆*If PBS and trypsin solution have been kept at 4℃, **can you use them as they are**?*

🔴 No, you should warm them up and keep them at room temperature.

🔴 いいえ，暖めて，室温に置いておきます．

🟢 What should you do if cells are not easily detached?

🟢 細胞が剥がれにくいとき，どのようにしますか？

🔴 You should use a cell scraper.

🔴 セルスクレイパーで剥がします．

2. 継代
…Subculturing

🟢 What is the reaction time for trypsin?

🟢 トリプシンはどれくらいの時間反応させますか？

🔴 It should be 2 to 3 minutes.

🔴 2〜3分です．

🟢 What should you do to stop the trypsin reaction?

🟢 トリプシンの反応を止めるときどうしますか？

🔴 You should add a medium containing FBS (fetal bovine serum).

🔴 FBS（牛胎仔血清）の入った培地を加えます．

● What ratio should you use when you seed 70% confluent cells?

● 70%コンフルエント（confluent）になった細胞をどれくらいの割合で播きますか？

🍡 Normally, you should **seed at a ratio of about 1 to 3**.

🍡 通常，1：3くらいの割合で播きます．

🌸 その他の表現法
◆ *Normally, the ratio should be about 1 to 3*.

Column テクニシャンとの関係

　アメリカではテクニシャンの分業制がはっきりしていて，実験手技に応じていろいろ手伝ってくれたり，助言してくれることがあります．ときに意見を求められることもあります．いろんな人がいますがこれらの人とうまく接することが実験をうまく進めるポイントになります．細胞培養を手伝ってくれたテクニシャンは培養室でいつも好みの音楽をハイボリュームで流していました．自分は好みでなくてもときには我慢して『いい曲だね．』とお世辞の1つも必要です．

Column スペイン語も？

　アメリカの病院（私が留学していたのは病院の隣に付属していた研究棟です）はボランティアの人を含め多くの方が支えています．毎日，ラボを掃除にきてくれる方がいましたが，彼等の中にはスペイン語圏からきている人も少なくありませんでした．毎日顔を合わせて話しているうちに，スペイン語の簡単な会話ができるようになりました．英語は一向に上達しないのにどうしてかと思っていましたが，あまり深く考えないようにしました．その気があればスペイン語のちょっとした会話も身につくかも？しれません．

Section Ⅱ：Let's speak! 〜実験中の英会話

Ⅱ・4 細胞培養をする

II●4 細胞培養をする

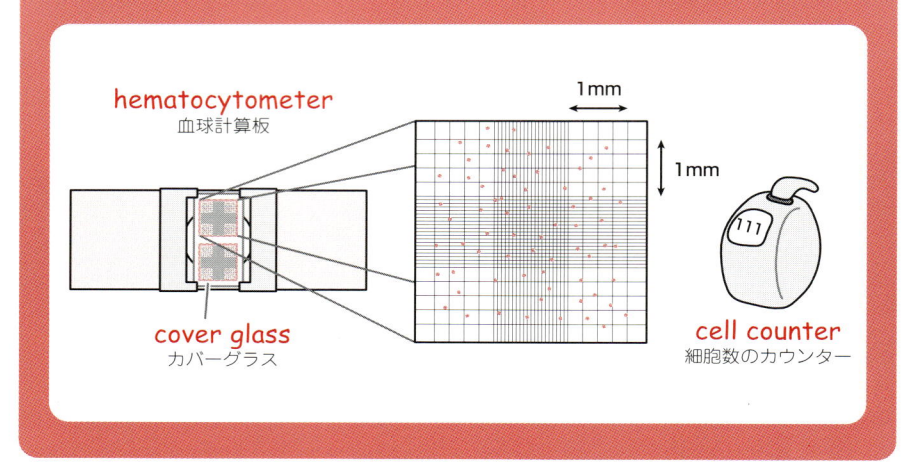

❺ 細胞数の測定

hematocytometer
血球計算板

1mm

1mm

cover glass
カバーグラス

cell counter
細胞数のカウンター

1. 細胞浮遊液の調整
…Preparation of cell suspension

🌑 How should you prepare a cell suspension?

🌑 細胞浮遊液はどのようにして調整しますか？

🔴 After you centrifuge the collected cells, you add a medium to the cell pellet.

🔴 集めた細胞を遠心後、そのペレットに培地を加えて調整します．

🌑 What should you be careful about when you prepare the cell suspension?

🌑 細胞浮遊液を調整するとき、どのようなことに注意しますか？

🔴 You need to **loosen the cells sufficiently** to make them even, leav-

🔴 十分に細胞をほぐして、細胞の塊が残らないよう

ing no blocks of cells.

🦐 その他の表現法
◆ *You need to <u>disperse the cells completely</u> to make them even, leaving no blocks of cells.*

🍀 What should you do if the cell suspension has blocks of cells?
🍀 細胞浮遊液に塊が残るとき，どのように処理しますか？

🍓 You should **pass them through** a sterilized filter.
🍓 滅菌済みのフィルターを通します．

🦐 その他の表現法
◆ *You should <u>filter them through</u> a sterilized filter.*

2．血球計算板での計測
…Determination of cell counts on the hemacytometer

🍀 What should you do when you place a cover glass over a hemacytometer?
🍀 血球計算板にカバーグラスをのせるときどのようにしますか？

🍓 You should do it in such a way that it forms a Newton ring.
🍓 ニュートンリングができるようにのせます．

🍀 What kind of reagent should you use to stain the cells?
🍀 細胞はどのような試薬で染めますか？

Section Ⅱ：Let's speak! 〜実験中の英会話

その他の表現法

◆ Which reagent should be used to stain the cells?

🐡 You should use trypan blue solution. 　　🐡 トリパンブルー（Trypan Blue）液で染めます．

🐡 If you can see 300 or more cells under the field of the microscope, should we continue to count them? 　　🐡 一視野に300以上の細胞が見えるとき，そのまま数えますか？

🐡 No, you should dilute the cells and count them again. 　　🐡 さらに希釈してから，もう一度数え直します．

Column　ラボミーティング

　実験もそこそこ進んでデータが出だしたとき，自分のデータを発表する機会が回ってきます．内容をあれやこれや考え，つたない英語でぼそぼそ発表をはじめたのですが，プロジェクターの音が半端ではない大きさで，1番はなれたところに座っているBossは最初，怪訝そうな顔をしていました．次の瞬間Bossの方へ目をやると完全に熟睡していました．自分のプレゼンテーションのまずさを痛感しました．日本人は一般にプレゼンテーションが苦手な人種です．会話で印象をよくするのはなかなか大変なのでせめてスライドやOHPはわかりやすいものにしたいところです．

II●4 細胞培養をする

6 細胞のクローニング

- cloning ring / クローニングリング
- culture dish / 培養皿
- 96-well culture plate / 96穴培養プレート

1. コロニーの選定
…Selection of colonies

● How should we mark the selected colonies?

● 選定したコロニーの目印をどのようにつけますか？

● **While they are still under the microscope**, you mark them from the back of the dish with a felt pen.

● 顕微鏡下で，dishの裏からマジックで印をつけます．

その他の表現法
◆ *Under the microscope, you mark them from the back of the dish with a felt pen.*

● How should you select the colonies?

● コロニーの選定はどのようにしますか？

🍑 You do so under a microscope, starting from a low magnification.

🍑 顕微鏡下で，弱拡から観察して選定します．

🍫 **Can you see** the colonies to be collected **with the naked eye**?

🍫 回収可能なコロニーは肉眼でも見えますか？

その他の表現法
◆ *Can you see* the colonies to be collected *without the aid of a microscope*?

🍑 Yes, you can.

🍑 肉眼でも観察可能です．

2. 細胞の回収
…Collection of cells

🍫 What kind of apparatus should you use to collect cells?

🍫 細胞の回収にはどんな器具を使用しますか？

🍑 You use a cloning ring.

🍑 クローニングリングを使用します．

🍫 How should you attach a cloning ring to a culture dish?

🍫 クローニングリングと培養dishの接着はどうしますか？

🍑 You should apply Vaseline over the area of the cloning ring.

🍑 クローニングリングの設置面にワセリンを塗ります．

🍫 **How can you collect cells** in the cloning ring?

🐛 その他の表現法
◆ *How can you retrieve cells* in the *cloning ring?*

🍊 You can use the yellow tip of a pipetman.

🍫 クローニングリング内の細胞をどのように回収しますか？

🍊 ピペットマンのイエローチップで回収します．

Column

慣れは恐い

　海外の留学先から学会発表をするとき，変なプレッシャーがあります．日本から海外の学会に演題発表に行ったときは，どうせ日本人だから少々，英語がまずくてもなんとかなるだろうと思っていました．しかし留学して，1年以上経っていると，日本からくる知り合いからは，留学しているのだから当然，質疑応答も無難にこなすだろうと見られているようです．実際は全くそんなことはありません．かえって英語が下手になったんじゃないかと思うくらいのときもあります．ただ，何が変わるかというと，外人恐怖症がなくなることです．少々内容がわからなくてもそれなりに対応できる自信はできます．こんなわたしでもそれなりに留学生活を送ってきたわけですから，留学前にあまり英会話のことで神経質になることはないと思います．でも，たわいのないことでも何か話しかけるように心がけておくことは必要かもしれません．

II・4 細胞培養をする

7 細胞の保存

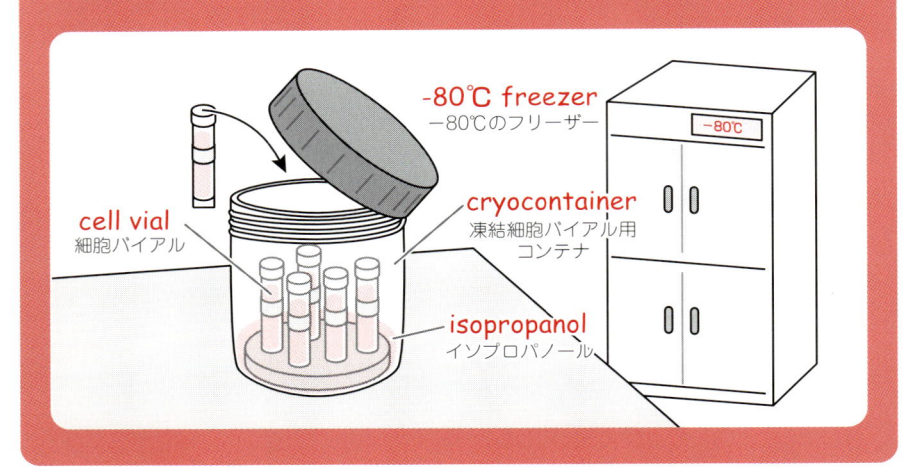

-80℃ freezer
−80℃のフリーザー

cell vial
細胞バイアル

cryocontainer
凍結細胞バイアル用コンテナ

isopropanol
イソプロパノール

1. 細胞懸濁液
…Cell suspension

🍫 What volume of cell suspension should you use?

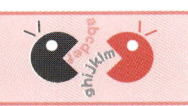

　　細胞懸濁液の容量はどれくらいにしますか？

🍓 It depends on the concentration of the cells.

　　細胞濃度によってかえます．

🍫 **What** does the cell suspension **consist of**?

　　細胞懸濁液の組成は何ですか？

　🔍 その他の表現法
　◆ <u>*What's the composition*</u> of the cell suspension?

🍓 It consists of a medium which has had DMSO added to it.

　　培地にDMSOを加えたものです．

その他の表現法
◆ *It is composed of a medium to which DMSO has been added.*

🍩 What is the concentration of fetal bovine serum in the medium?

🐟 It is normally 20%, but can be higher depending on the cells.

培地に含まれる牛胎仔血清の割合はどれくらいですか？

通常，20％ですが，細胞によってはさらに高濃度にします．

---

2. 凍結
…*Freezing*

🍩 Should we place the cells in a －80℃ freezer immediately?

🐟 No, the **temperature should be lowered gradually**.

その他の表現法
◆ *No, you should <u>reduce the temperature gradually</u>.*

細胞はすぐに－80℃に入れますか？

いいえ，徐々に温度を下げていきます．

🍩 How should you do that?

🐟 We should first place the cells in an isopropanol-filled cryocontainer

そのためにどんな方法を使いますか？

イソプロパノールを入れた細胞保存コンテナに入

and then transfer them to the −80℃ freezer.	れてから−80℃に移します．
🔴 Can we store them at −80℃ for a long period of time?	🔴 −80℃で長期間保存できますか？
🔴 They should usually be transferred to a liquid nitrogen tank depending on the cells.	🔴 細胞によりますが，長期間保存には通常液体窒素に移します．

Column 論文の準備…でも進まない

　データも出そろい，自分の成果を論文にまとめることができそうになってきたら論文の作成に取りかかるわけですが，この際，論文用に画像を調整したり，高品質の写真でプリントアウトする必要が出てきます．普段使い慣れている個人用のcomputerはともかく，上記の作業は大抵，ラボにある共用のcomputerやその周辺機器を使って行います．これらの機器の扱いに強い方は問題ありませんが，私は結構苦労しました．高画質で取り込んだはずの写真が，実際にはかなり質が悪かったり，ひどいときはfileそのものが開けないこともありました．日本のcomputerの環境（OS）とアメリカのものとがうまく噛み合わなかったり，MacとWindowsの互換性の問題で悩んだこともありました．普段実験に集中しているときはなかなか時間がありませんが，できればラボのcomputerや周辺機器の扱いにちょっとでも慣れておくといざというときに役立つと思います．

II●4 細胞培養をする

🎱 培養実験を終えて

```
sink              plastic bottle
シンク             for waste liquid
                  廃液用の
                  プラスチックボトル

         culture bottle        waste
         培養ビン                liquid
```

1. 培養器具の処理
…Disposal of cell culture equipment

🎨 What should you do with a pipetman after we have used it?

🎨 使用後のピペットマンはどうしますか？

🍡 You should **wipe it gently with cotton soaked in alcohol** and store it away from the clean bench.

🍡 アルコール綿で軽く拭って，クリーンベンチの外で保管します．

🎨/その他の表現法
◆ *Wipe it lightly with cotton soaked in alcohol and store it away from the clean bench.*

🎨 What should you do about the clean bench?

🎨 クリーンベンチはどうしますか？

- You should wipe the surface with cotton soaked in alcohol and leave the UV lamp turned on.
- アルコール綿で表面をふいて，紫外線ランプを点灯しておきます．

- What should you do about pipettes?
- ピペットはどうしますか？

- You should wash and dry glassware and use it again after you have **sterilized it with dry heat**.
- ガラス製品は，洗浄して乾燥後，乾熱滅菌をかけて再利用します．

  🐛 その他の表現法
  ◆ *You should wash and dry glassware and use it again after you <u>sterilize it in a dry heat oven</u>.*

2. 培養液の処理
…Disposal of culture liquid

- What should you do about culture liquid in an aspiration bottle?
- 吸引ビンの培養液はどのようにしますか？

- You should add hypochlorous acid and dispose of it as an alkaline solution.
- 次亜塩素酸を加えてアルカリ液として処理します．

- What should you do about the remaining culture liquid?
- 残った培養液はどうしますか？

🔵 You should store it at 4℃.

🔵 4℃で保存しておきます．

🟡 What should you do about the remaining fetal bovine serum?

🟡 残った牛胎仔血清はどうしますか？

🔵 You should **divide it up into separate** tubes and store them at －20℃.

🔵 分注して，－20℃で保存します．

その他の表現法
◆ You should **make aliquots** and store them at －20℃.

ラボの廊下

　留学先のラボの廊下には数々の業績や，学会の写真が飾られている．その一角に，代々のラボのメンバーの白黒写真が額に入って飾られている．日本人としては少し変な気分にもなりますが，単に業績の論文だけが貼り出されている場合と違い，ラボの全メンバーを思いやるBossの気持ちが感じられました．学会前，presentationの練習のため，つたない英語の私を相手に遅くまでつきあってくれたBoss，結構辛口のことも言われましたがいずれも研究者へのよきアドバイスとして私の心に残っています．そのBossは一昨年10月の終わりに突然他界されてしまいました．学会での私たちのpresentationの成果を見てもらう直前でした．留学後，いろいろな理由でラボを移ったりされる方もたくさんおられます．また私たちのような経験をする研究者たちもいるのです．

細胞培養をする

Section III

4コマ漫画でシミュレーション
～実験結果を報告しよう！

登場人物紹介

Director：研究室のボス

Cathy：留学先の研究室の先輩

Satoshi：留学したての日本人研究者

❶ インサートの確認
制限酵素でDNAが切れない
【Checking the insert : unable to cut DNA with a restriction enzyme】

The subcloned DNA is treated with the restriction enzyme.
Digested DNA is run onto the gel.

制限酵素処理したDNAをゲルに流します．

The gel is photographed after electrophoresis, but no target DNAs have been cut at all. The Director is being consulted to clarify the cause.

電気泳動後，ゲルの写真を撮りましたが，目的のDNAは全く切れていませんでした．そこで，原因を明らかにするためボスとディスカッションしました．

どうして制限酵素処理でDNAが切れないのでしょうか？

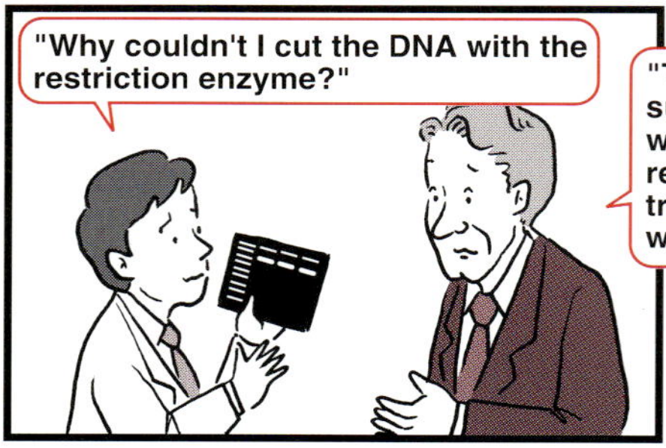

"Why couldn't I cut the DNA with the restriction enzyme?"

"The target DNA may not have been subcloned properly, or maybe there wasn't enough of a restriction enzyme reaction. Anyway, first, I suggest you try another enzyme, and if that doesn't work, then try subcloning again."

目的のDNAが正しくサブクローニングされていないか，制限酵素反応が十分でない可能性がある．とりあえず，別の制限酵素を試してみて，それでもだめなら，もう一度サブクローニングのやり直しだ！

"I'll try another restriction enzyme !"

別の制限酵素で酵素反応を試してみよう！

110 バイオ実験 英語でトライ！

2 インサートの確認
制限酵素で予想外の断片が出る
【Checking the insert : unexpected fragments appear with restriction enzyme digestion】

The subcloned DNA is treated with the restriction enzyme.
Digested DNA is run onto the gel.

サブクローニングしたDNAを制限酵素で処理します。
制限酵素処理したDNAをゲルに流します。

2.2 kb ··· 2.8kb
0.8 kb ··· 0.2kb

The gel is photographed after electrophoresis, but what is seen is different in size from the target DNA. The Director is being consulted to clarify the cause.

電気泳動後、ゲルの写真を撮ってみましたが、目的のDNAのサイズと異なっていました。そこで、原因を明らかにするためボスとディスカッションしました。

今回はバンドが2本でましたが、目的のサイズと異なっていました！

"I had two bands this time, but the sizes were different from the target DNA."

"OK. So, why don't you check the map of the target DNA? With this subcloning, maybe you inserted the DNA in the opposite orientation. (OR maybe you inserted the DNA the wrong way round.) You can confirm this using another restriction enzyme."

目的のDNAのマップをチェックしてみてはどうかな？
今回のサブクローニングではインサートDNAが反転して組み込まれる可能性もある。別の制限酵素で反応させてみて確かめなさい。

"O.K. I'll try another restriction enzyme and confirm the results by checking the DNA map."

他の制限酵素で反応させて、結果をDNAマップと照らし合わせて確認します。

Section Ⅲ : 4コマ漫画でシミュレーション〜実験結果を報告しよう！　111

③ 形質転換の結果
コロニーが形成されない
【Results of transformation : colonies are not formed】

The transformed *E. coli* cells are incubated in a test tube using a shaker.

形質転換後の大腸菌（*E. coli*）をシェーカーを用いて，試験管内で増殖させます．

After incubation, *E. coli* cells are spread over a solid medium.

増殖後の大腸菌（*E. coli*）を固形培地に塗布します．

"Wow! There's not a single colony on the plate! I wonder if I added too much antibiotic."

あれっ！プレートにコロニーが全然はえていない．固形培地に添加した抗生物質を入れすぎたのかな？

Cathy: "Where's the LB medium you used?"

Satoshi: "I used the one in the top left hand corner of the fridge."

Cathy: "Oh, I made that a year ago! Sorry, I guess I forgot to throw it away!"

Satoshi: "Oh well, I guess I'd better take the time to make it myself from now on."

キャシー： どこに置いてあったLB培地を使ったの？

サトシ： 冷蔵庫の左上隅に置いてあったLB培地を使わせてもらったんだけど…．

キャシー： あれは，1年前に私がつくったものです！捨てるのを忘れていました！

サトシ： 今度からは，急がずに自分でつくるよ…．

112　バイオ実験 英語でトライ！

④ 形質転換の結果
セルフライゲーションのコロニーのみ形成
【Results of transformation: only colonies of self-ligated products formed】

The transformed *E. coli* cells are spread over a solid medium.

形質転換後の大腸菌を固形培地に塗布します.

On the following day, colonies are collected from the solid medium and then grown in a liquid medium.

翌日,固形培地にできたコロニーをとり,液体培地内で増殖させます.

The DNA extracted by the alkaline prep method is electrophoresed.

アルカリプレップ法により抽出したDNAを電気泳動する.

Satoshi: "I thought there would be two bands, but there's only one. Yet the size is almost identical to the original vector!"

Director: "Mmm... Sounds like a case of self-ligation to me."

Satoshi: "How did that happen?"

Director: "Well, after you cut the vector with the restriction enzyme, did you dephosphorylate the end of the vector DNA with alkaline phosphatase?"

Satoshi: "Ah! No, I didn't. Now I know why!"

サトシ: 予想では2本出るはずのバンドが,1本しか出ません.しかももともとのベクターのサイズとほとんど同じです.
ボス: セルフライゲーションを起こしているね.
サトシ: どうしてこんな結果になったのですか?
ボス: ベクターを制限酵素で切断後,ベクターのDNA末端をアルカリホスファターゼによって脱リン酸化したかね?
サトシ: しませんでした.それで原因がわかりました.

Ⅲ 実験結果の報告

Section Ⅲ : 4コマ漫画でシミュレーション〜実験結果を報告しよう! 113

⑤ PCR産物の確認
増幅バンドが検出されない
【Checking the PCR products : no amplification bands are detected】

The target DNA is amplified in the PCR machine.

PCRの機器で目的のDNAを増幅します.

The PCR product is electrophoresed.

PCR産物を電気泳動します.

The stained gel is observed and photographed, but there are no bands other than those of the molecular weight markers.

染色後のゲルを観察し,写真を撮りましたが分子量マーカー以外全くバンドを認めませんでした.

Satoshi: "Why wasn't the target band amplified?"

Director: "Perhaps the annealing temperature was too high."

Satoshi: "But I did it at 60°C."

Director: "Well, I suggest you use a slightly lower temperature and try doing the PCR reaction again."

サトシ: 目的のバンドが増幅されなかったのはどうしてでしょうか?
ボス: アニーリングの温度が高すぎたのかもしれない.
サトシ: 今回,60℃で行いました.
ボス: もう少しアニーリングの温度を下げて,もう一度PCR反応を行ってみなさい.

6 PCR産物の確認
非特異的バンドが多い
【Checking the PCR products : many non-specific bands are detected】

"Who's using the PCR?"

"Oh, I am. But I'll be finished by 3 o'clock."

サトシ: このPCRは誰が使用中ですか？
キャシー: 私が使っています。3時には終わります。

いいえ、入ってないわよ。終わったらあなたに知らせるわ。

"No, I don't think so. I'll let you know when I've finished."

"OK. I'd like to use it next. Has anyone reserved it?"

次に使いたいんだけど、他に予約は入ってますか？

PCR is performed, followed by electrophoresis. The stained gel is observed and photographed, but there are numerous nonspecific bands.

PCRを行い電気泳動を行う。染色後のゲルを観察し、写真を撮りましたが多数の非特異的バンドを認めました。

Satoshi: "I did a PCR and found a lot of nonspecific bands."

Director: "The quality of DNA may have been bad, or there may have been some problems in the primer design."

Satoshi: "Well, I annealed it at 55°C."

Director: "Well, in that case, I think the annealing temperature was probably OK, so I suggest you check the DNA absorbance readings and the GC content of the primer."

サトシ: PCRの結果、多数の非特異的バンドが検出されました…。

ボス: DNAの質が悪かったか、プライマーのデザインに問題があったかもしれないね。

サトシ: アニーリングの温度は55℃で行いました。

ボス: アニーリングの温度に問題はないと思うので、DNAの吸光度の値と、プライマーのGCコンテンツをチェックしてみなさい。

Section Ⅲ : 4コマ漫画でシミュレーション〜実験結果を報告しよう！　115

7 ノーザンブロッティングの結果
バンドが検出されない
【Results of Northern blotting : no bands are detected】

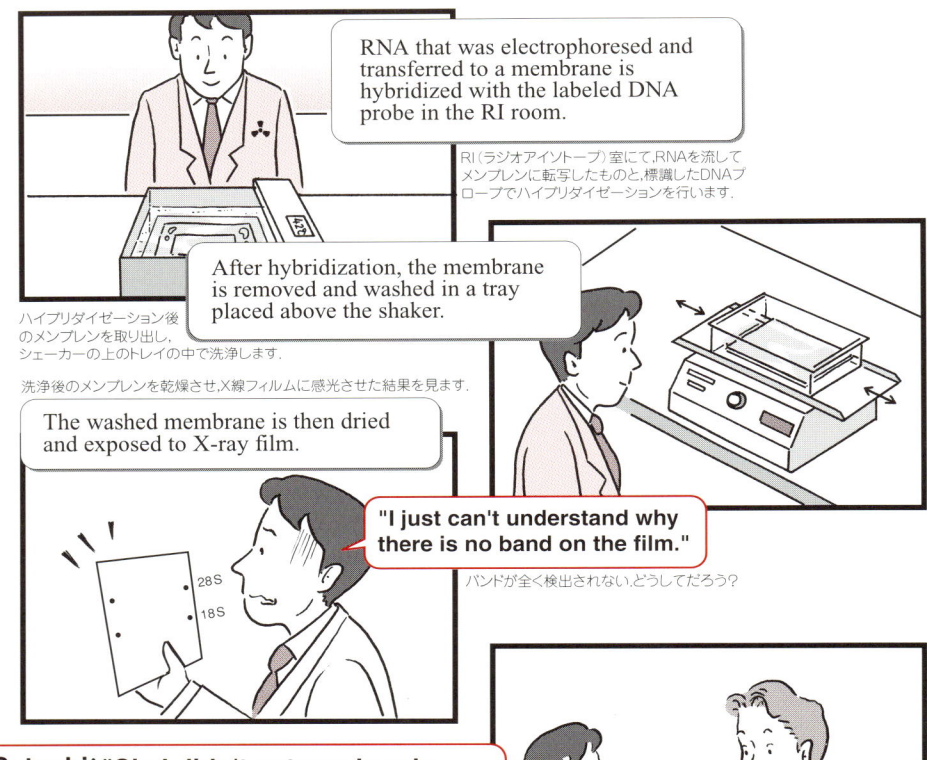

RNA that was electrophoresed and transferred to a membrane is hybridized with the labeled DNA probe in the RI room.

RI（ラジオアイソトープ）室にて,RNAを流してメンブレンに転写したものと,標識したDNAプローブでハイブリダイゼーションを行います．

After hybridization, the membrane is removed and washed in a tray placed above the shaker.

ハイブリダイゼーション後のメンブレンを取り出し,シェーカーの上のトレイの中で洗浄します．

洗浄後のメンブレンを乾燥させ,X線フィルムに感光させた結果を見ます．

The washed membrane is then dried and exposed to X-ray film.

"I just can't understand why there is no band on the film."

バンドが全く検出されない,どうしてだろう？

Satoshi："Sir, I didn't get any bands after Northern blotting."

Director："Maybe either the quality of the electrophoresed RNA was poor, or the level of expression of the target RNA was below the detectable range."

Satoshi："What else can I do if the RNA expression is low?

Director："How about checking it with an RNAase protection assay or trying the PCR method?"

サトシ：ノーザンブロッティングの結果,バンドが全く検出されないのですが…．

ボス：電気泳動したRNAの質が悪かったか,目的のRNAの発現量が測定感度以下だったかもしれない．

サトシ：RNAの発現量が少ないとすると,他にどんな確認方法がありますか？

ボス：RNAaseプロテクションアッセイかPCR法で確認してみてはどうだ．

❽ ノーザンブロッティングの結果
バックグラウンドが高い
【Results of Northern blotting : high background】

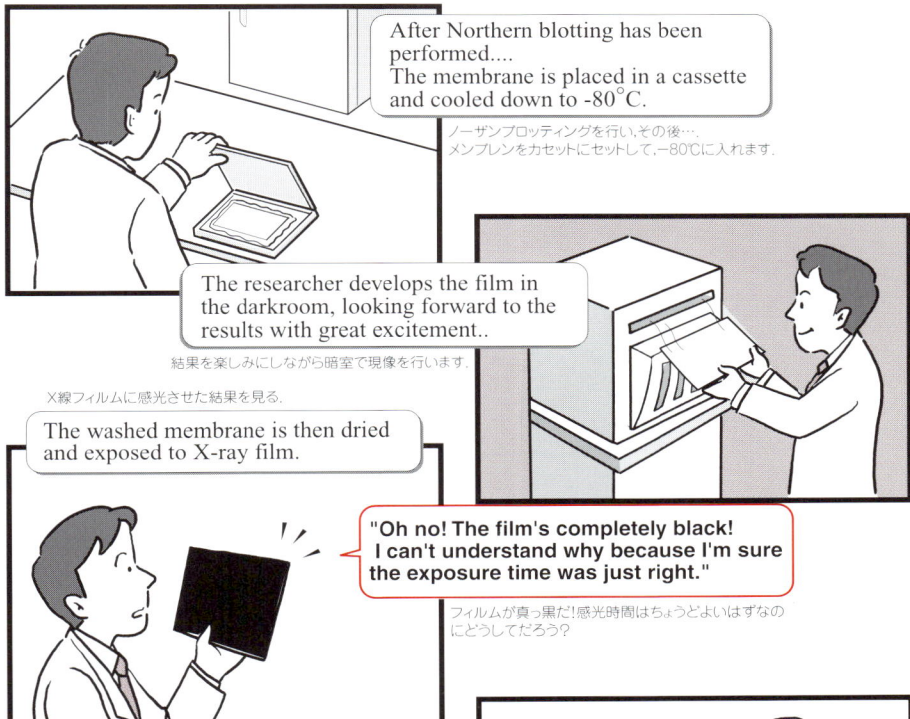

After Northern blotting has been performed....
The membrane is placed in a cassette and cooled down to -80°C.

ノーザンブロッティングを行い,その後….
メンブレンをカセットにセットして,-80℃に入れます.

The researcher develops the film in the darkroom, looking forward to the results with great excitement..

結果を楽しみにしながら暗室で現像を行います.

X線フィルムに感光させた結果を見る.

The washed membrane is then dried and exposed to X-ray film.

"Oh no! The film's completely black! I can't understand why because I'm sure the exposure time was just right."

フィルムが真っ黒だ!感光時間はちょうどよいはずなのにどうしてだろう?

Satoshi: "Sir, I did Northern blotting, but the film is completely black."

Director: "Did you denature the DNA probe to make it single-stranded before you performed hybridization?"

Satoshi: "Yes, I did. I didn't forget that!"

Director: "Well then, did you remove the free RI after you labeled the DNA probe?"

Satoshi: "Oh no, I completely forgot about that! That could well be the reason, then."

サトシ: ノーザンブロッティングの結果,フィルムが真っ黒になったのですが…
ボス: ハイブリダイゼーションの前に,DNAプローブを1本鎖に変性させる操作を行ったかね?
サトシ: それは忘れずに行いました!
ボス: それでは,DNAプローブの標識後,遊離のRI(ラジオアイソトープ)を除去したかね?
サトシ: あっ!それを忘れました.それが原因かもしれませんね.

Section Ⅲ : 4コマ漫画でシミュレーション〜実験結果を報告しよう! 117

9 ウエスタンブロッティングの結果
バンドが検出されない
【Results of Western blotting : no bands are detected】

The protein is run onto a gel using electrophoresis apparatus.

電気泳動装置を用いてタンパク質をゲルに流します．

泳動後のゲルをメンブレンに転写します．

After electrophoresis, the gel is transferred to a membrane.

感光時間を15分にしたけれど，バンドが全く検出されない，どうしてだろう？

"The exposure time was 15 minutes, but there's no band. I don't understand why."

The membrane undergoes antibody reaction, is stained with a chemiluminescent reagent, and exposed to X-ray film. The film is now being observed.

抗体反応を行い，化学蛍光発色後感光したX線フィルムを見て….

Satoshi: "Sir, I did Western blotting, but I didn't get any bands and I'm wondering why."

Director: "Well, either the concentration of the primary antibody could have been too low, or the secondary antibody may not have been suitable for the primary antibody."

Satoshi: " I see. Right, I'll try again using the antibody to the protein, which was expressed at high levels in the cells I extracted this protein from.

Director: "Right. Then, you'll know whether the protein you extracted had a problem or not, right?"

サトシ: ウエスタンブロッティングの結果，バンドが全く検出されませんでした．どうしてでしょうか？

ボス: 1次抗体の希釈率が薄すぎたか，2次抗体の種類が1次抗体とあってなかったかもしれない．

サトシ: このタンパク質を抽出してきた細胞で大量に発現しているタンパク質の抗体でもう一度試してみます．

ボス: それで，抽出してきたタンパク質の質に問題がなかったかどうか判断できるね．

10 ウエスタンブロッティングの結果
バックグラウンドが高い
【Results of Western blotting : high background】

I guess I'll start making the gel all over again.

At 100 volts the bands weren't very clear, so I'll run it at 30 volts overnight.

"The exposure time was only 30 seconds, but the film's completely black."

The membrane undergoes antibody reaction, is stained with a chemiluminescent reagent, and exposed to X-ray film. The film is now being observed.

Satoshi: "Sir, I did Western blotting and found the film was totally black. Why would that be?"

Director: "Maybe the concentration of the blocking reagent in the blocking buffer you used before the antibody reaction was too low. I suggest you try using higher concentrations."

Satoshi: "I used skimmed milk as a blocking reagent."

Director: "Well, it may be a good idea to try bovine serum albumin or gelatin."

INDEX
キーワードで探す英語表現

※欧文後の数字は本文中のページを示しています.

単 位	
bp	base pair
cfu	colony forming unit
Da	dalton
μg	microgram
ml	milliliter
mM	millimolar concentration
moi	multiplicity of infection
O.D.	optical density
pfu	plaque forming unit
rpm	revolutions per minute
U	unit
w/v	weight per volume

abc	
CBB染色	CBB staining (21)
Colloidal Gold染色	Colloidal Gold staining (21)
DEPC添加水	DEPC-treated water (18)
DNA-タンパク質複合体	DNA-protein complex (73)
PCR反応	PCR reaction (59)
Ponceau S染色	Ponceau S staining (21)
RIの入ったバイアル	RI vial (15)
SDS-PAGE用ゲル	SDS-PAGE (sodium dodecylsulfate-polyacrylamide gel electrophoresis) (20)
SYBR-Gold染色	SYBR-Gold staining (21)
SYBR-Green染色	SYBR-Green staining (21)
Taqポリメラーゼ	Taq polymerase (59)
X線フィルム	X-ray film (15)
X線フィルムに感光させる	expose it to X-ray film (64) X-ray it (64)

ア 行

アガロースゲル	agarose gel (20)
アスピレーター	aspirator (16), (84)
値は正確ですか？	Are the values you obtain from prestain markers used in the SDS-PAGE method accurate? (72) Are the figures you obtain from prestain markers used in the SDS-PAGE method exact? (72)
あまり長くインキュベーションし過ぎないように気をつけます	You should take care not to over-incubate the liquid media. (44) You should be careful not to incubate the liquid media for too long. (44) You need to take care not to incubate the liquid media too much. (44)
アルカリ	alkali (32)
アルカリ液として	as an alkaline solution (106)
アルカリ溶液	alkaline solution (46)
アルコール綿で軽く拭って	You should wipe it gently with cotton soaked in alcohol 〜 (105) Wipe it lightly with cotton soaked in alcohol 〜 (105)
（pHメーターで）合わせる	You can adjust the pH with an acid or alkali using a pH meter. (30) You can match the pH to the required value with an acid or alkali using a pH meter. (31)
椅子	stool (16), (17)
（今度からは)急がずに自分でつくるよ	I guess I'd better take the time to make it myself from now on. (112)
イソプロパノール	isopropanol (102)
一昼夜	overnight (54)
１本ずつ	one by one (88)
（37℃恒温器の中に，30分ほど）入れておきます	You incubate them in an incubator at 37℃ for 30 minutes. (41) You place them in an incubator and incubate at 37℃ for 30 minutes. (41)
インサートDNA	insert DNA (52)
内扉	inner door (17)
影響を受けにくい	less affected (72) have less of an effect (72)
（30ボルトでオーバーナイトで）泳動しよう	I'll run it at 30 volts overnight. (119)
泳動槽	electrophoresis tank (63), (71)
泳動方向	electrophoresis direction (63)
液体窒素タンク	liquid nitrogen tank (22), (87)
液体培地	liquid media (LB media) (46)
液体流出口	ejection site (12)
エチジウムブロマイド染色	ethidium bromide staining (21)
エレクトロポレーションシステム	electroporation apparatus (57)
エレクトロポレーション法	electroporation (57)
塩基数	number of the base (66)
遠心濃縮機	centrifuge desiccator (14)
オートクレーブ	autoclave (17)

オートクレーブ水	autoclaved water （18）
オートシークエンサー	autosequencer （63）
オートシークエンサーの結果のグラフ	electropherogram （66）
オーバーナイト	overnight, O/N （54）
（3時には）終わるよ	I'll be finished by 3 o'clock. （115）
温度，炭酸濃度調節パネル	temperature and CO_2 concentration indicator （17）
温度調節つまみ	temperature control （13），（16），（17）
温度調節ボタン	temperature control （14）
温度表示窓	temperature indicator （13），（17）
（37℃の）温浴槽	water bath (37℃) （87）
オンラインでデータを取りだす	get data on-line （67） go on-line and obtain data （67）

カ　行

ガイガーカウンター	Geiger counter （15）
（細胞をどのように）回収しますか？	How can you collect cells （101） How can you retrieve cells （101）
（コンピュータソフトを使用して）解析します	You analyze them using computer software. （65） You use computer software. （65）
解析はどのようにして行いますか？	How do you analyze ～ ?（64）
回転軸	rotation axis （17）
回転数表示窓	rotation speed indicator （17）
回転台	rotator （14）
回転数調節つまみ	rotation control （12），（13），（14），（17）
核酸	nucleic acid （21）
（～で)確認してみてはどうだ	How about checking it with ～? （116）
ガスバーナー	Bunsen burner （16），（17），（93）
ガス管	gas tube （17）
ガス流出口	gas tap （17）
カセット（X線フィルム）	X-ray film cassette （15）
カバーグラス	cover glass （96）
ガラスピペット	glass pipette （46）
乾燥器	oven （13）
簡単な操作で一度に	in one simple operation （58）
乾熱滅菌をかけて	sterilized it with dry heat （106） sterilize it in a dry heat oven （106）
簡便に行える	It can be performed easily ～? （58） It's easy to do ～? （58）
関連，～の間	between A, B and C? （40） among A, B and C? （40）
（～のような)器具を用意する	The equipment comprises ～（70）
既知の遺伝子	known gene （67） recognized gene （67）
気泡が入らないよう	not to let air bubbles get in （63） avoid letting air bubbles in （63）
逆向きに挿入されている	in the opposite or "reverse" orientation （56）

	the wrong way round (56)
	reversed insertion (52)
キャップ	cap (12)
吸引管接続口	connector to aspirator (16)
吸引装置	vacuum unit (14)
給気プレフィルター	air filter (17)
(カラムのメンブレンにDNAを) 吸着させ，最後に抽出してくるタイプのものがあります	There is one kit where **DNA is adsorbed to a column membrane** and then eluted. (47)
	There are several types. For example, there is one kit where **DNA is adsorbed to a column membrane** and then eluted. (47)
キュベット	cuvette (57)
教授	professor, principal investigator (P.I.) (25)
切り出し	cut out (53)
銀染色	Silver staining (21)
クライオスタット	cryostat (78)
(〜と) 比べてどうですか？	**How does** RNAse protection assay **compare to** 〜? (62)
	How does RNAse protection assay **measure up to** 〜? (62)
クリーンベンチ	clean bench (16), (93)
クリップ	clip (69)
クローニングリング	cloning ring (99)
蛍光顕微鏡	fluorescence microscope (15)
蛍光灯	fluorescent light (16)
劇物	deleterious substance (30)
血球計算板	hemacytometer (16), (96)
(〜の)欠点は何ですか？	What are the disadvantages of 〜? (66)
	What is the drawback of 〜? (66)
ゲル	gel (21), (49), (71)
ゲルが大きい	a large amount of gel (63)
	gel is thick (63)
ゲルシフトアッセイ	electro- mobility shift assay (EMSA) (73)
ゲルに流す	run onto the gel (110)
ゲルづくりからやり直しだ！	I guess I'll start making the gel all over again. (119)
ゲル作製台	gel preparation table (71)
ゲル作製枠	gel cassette (14)
ゲル台, ゲル板	gel plate (15). (63)
(それが) 原因かもしれませんね	**That could well be the reason, then.** (117)
(それで) 原因がわかりました	**Now I know why!** (113)
原因を明らかにするためボスとディスカッションしました	The Director is being consulted **to clarify the cause**. (110)
研究室の構成	laboratory hierarchy (25)
検出限界以下	below the detectable range (79)
検出不可能	not detectable (66)
顕微鏡下で	While they are still under the microscope (99)
	Under the microscope (99)
顕微鏡で観察する	**observe** cultured cells under a microscope (91)
	take a look at cultured cells **with a microscope** (91)

恒温槽	water bath (16)
講師	assistant professor (25)
考慮して	taking into consideration (59)
(目的のDNAのサイズと)異なっていました	What is seen **is different in size** from the target DNA. (111)
コネクター	connector (12)
コロニー	colony (40)

サ 行

(インサートDNAの) サイズのチェックはどのようにして行いますか？	What method is used to **check the size of** the insert DNA? (55) How do you **check the size of** the insert DNA? (55)
細胞数のカウンター	cell counter (96)
細胞毒性を持った物質	cytotoxic substances (89)
細胞バイアル	cell vial (102)
細胞浮遊液	cell suspension (96)
細胞をほぐす	loosen the cells sufficiently (96) disperse the cells completely (97)
再利用できますか？	**Can you use** a reagent bottle that has contained liquid medium again? (84) **Can you recycle** a reagent bottle that has contained liquid medium? (85)
殺菌灯	UV light (16), (17)
酸	acid (32)
サンガー法	Sanger method (63)
三角フラスコ	conical flask (13), (46)
産業廃棄物として	as hazardous industrial waste (32)
サンプルをウェルに入れるときに気をつけることは何ですか？	What do you need to be careful about when you **apply samples**? (50) What do you need to be careful about when you **put the samples in the wells**? (50) When **loading a gel**, what precautions should you take? (50)
シークエンサー	sequencer (63)
シェーカー	shaker (17), (43)
紫外線照射装置	transilluminator (15)
紫外線防護メガネ	UV safety glasses (15)
試験管	test tube (43)
実験計画をたてる	establish a plan (26)
自動現像機	automatic developer (15)
(ウェルの端を壊さない)〜しないように確実にサンプルをウェルに入れることです	You should **take care not to** break the edge of a well, and you should avoid letting the samples leak out. (51) You should **take care not to** puncture the edge of a well with the tip of the pipette and you should also be careful not to overfill it, as this will cause the sample to leak out. (51)
シャーレ	petri dish (17)
シャーレ設置面	petri dish table (17)
蛇口	tap (13)
試薬のグレード	grade of reagents (27)

試薬名	name of reagents (27)
遮光する	shield it from the light (31)
	protect it from the light (31)
遮へい板	shield (15)
(〜の) 種類によって	vary depending on what kind of (44)
	differ according to the type of (45)
使用期限	expiration date (31)
	How long can you use this reagent for? (31)
上下に強くふる	shaking vigorously (19)
上下反対にする	the plate upside down (42)
	invert the plate(42)
	turn the plate upside down (42)
使用後の電気泳動槽はどうしますか？	What should you do with the electrophoresis tank when you've finished with it ?(34)
情報を知る	access information (67)
蒸留水	distilled water (18)
少量培養	small scale amplification (43)
助教授	associate professor (25)
助手	research assistant (25)
徐々に温度を下げる	temperature should be lowered gradually (103)
	reduce the temperature gradually (103)
試料	sample (63), (69)
シンク（流し）	sink (13), (105)
シンクに流す（捨てる）	throw this waste water away in the sink? (32)
	dispose of this waste water in the sink? (32)
振動調節つまみ	vibrator unit/cup (12)
振動板	vibration control (12)
推察されます	One might surmise 〜 (91)
	This could imply that 〜 (91)
水層	water layer (38)
スイッチを入れるだけですぐ使えますか？	Can you use the automatic processor immediately after switching it on? (28)
	Can you use the automatic processor straight after you switch it on? (28)
水道水	tap water (18)
スクレイパー	cell scraper (93)
スターラー	stirrer (magnetic stirrer) (13)
スターラー用のバー	stirring bar (13)
素手で	with bare hands (88)
	without wearing gloves (88)
素早く37℃で融解する	thaw them quickly (87)
	defrost them promptly (87)
	unfreeze them rapidly (87)
スプレッダー	spreader (17)
(〜を) 正確に反映する	it reflects (the molecular weight of single-stranded RNA) more accurately. (61)
	it shows (the molecular weight of single-stranded RNA) more precisely. (61)

制限酵素	restriction enzyme (53)
制限酵素サイト	restriction enzymatic site (52)
制限酵素処理	restriction reaction (53), (55)
制限酵素処理をする	digest (55)
	perform restriction reaction (56)
製造会社名	name of manufacturer (27)
接眼レンズ	eyepiece lens (90)
洗浄用水	detergent solution (18)
相互作用を解析することに	It **analyzes the interaction** of DNA with a transcription factor. (74)
より測定します	It **examines how** transcription factors **interact with** DNA. (74)
速度調節つまみ	speed control (12)
組成は何ですか？	**What** does the cell suspension **consist of**? (102)
	What's the composition of the cell suspension? (102)
～の組成です（培地にDMSO	It **consists of** a medium which has had DMSO added to it. (102)
を加えたものです）	It **is composed of** a medium to which DMSO has been added. (103)
注ぎ口	lip (13)
袖まくり	**roll up** your sleeves (25)
外扉	outer door (17)

タ 行

大学院生	graduate student (25)
大腸菌塗布ターンテーブル	inoculation turntable (17)
大腸菌ペレット	bacterial pellet (46)
タイマー	timer (12), (13), (14)
大量培養	large scale amplification (43)
正しいサブクローニング	correct subcloning (52)
脱イオン化蒸留水	ddH$_2$O-distilled deionized water (=double distilled water) (18)
(～を)試してみてもいいかも	It may be a good idea to try ～ (119)
しれない	
(このPCRは) 誰が使用中？	Who's using the PCR？ (115)
短時間で	in a short period of time (48)
短時間に解析する	**analyze** several samples **in a short time** (64)
	study several samples **in a short period of time** (64)
	examine several samples **in a short space of time** (64)
タンパク質	protein (21)
タンパク質の同定	identify proteins (76)
	ascertain which proteins interact (76)
(～と～の)	What's the difference **between** liquid media and agar media? (40)
違いはなんですか？	**In what way do** liquid media and agar media **differ**? (40)
チップ	tip (12), (49)
チップ交換ボタン	tip replacement button (12)
中間層（変性したタンパク質）	intermediate layer (38)
チューブ, 試験管	test tube (12)
(1.5mlの) チューブ	(1.5ml) tube (46), (75)
（エッペンドルフチューブ）	
チューブたて	tube rack (12)
チューブ卓上遠心機	bench-top centrifuge (12)
長期間保存	for a long period of time (104)

超純水	milli-Q water (=nanopure water =ultrapure water) (18)
使い捨てピペット	disposable pipette (12)
低倍率から	from a low magnification (92)
デジタルカメラ	digital camera (90)
手袋	gloves (16)
電圧調節つまみ	voltage control (14)
電気泳動槽	electrophoresis tank (14)
電気泳動装置	electrophoresis apparatus (14), (49)
電極線接続部	electric cord connector (14)
電源	power supply (49), (71)
電源装置	power supply (14)
転写装置	transcription apparatus (71)
電子レンジ	microwave oven (14)
転倒混和	flipping upside down (19)
倒立型顕微鏡	inverse microscope (16), (90)
電動ピペッター	auto pipetor (12)
電動ポンプ(アスピレーター用)	automatic pump (aspirator) (16)
天秤	balance (13)
天秤皿	balance dish (13)
塗布します	spread over (112)
電流調節つまみ	mA control (14)
(目的のDNAのマップをチェックしてみては)どうかな？	**Why don't you** check the map of the target DNA? (111)
どうしてこんな結果になったのですか？	How did that happen? (113)
(感光時間はちょうど良いはずなのに)どうしてだろう	**I can't understand why** because I'm sure the exposure time was just right. (117)
(〜の記録は)どうしますか？	How should you keep a record 〜 (33) What records should you keep 〜 (33)
透析膜	dialysis membrane (69)
透析用バッファー	dialysis buffer (69)
(〜の) 特徴は何ですか？	What is the advantage of 〜 ? (58)
毒物	poisonous substance (30)
取っ手	grip (12)
どのような器具を用いますか？	What kind of equipment is used 〜 ? (70) What type of apparatus is used 〜 ? (70)
どのようなことが考えられますか？	What would it mean 〜 ? (91)
(DNA溶液と塩化カルシウム溶液を混和するとき)どのようなことに気をつけますか？	**How do you** mix DNA solution and Calcium Chloride solution? (57) **What do you need to be careful about** when you mix DNA solution and Calcium Chloride solution? (57)
(細胞は) どのような試薬で染めますか？	**What kind of reagent** should you use to stain the cells? (97) **Which reagent** should be used to stain the cells? (98)
どのような条件で行いますか？	How do you 〜 ? (43) Under what conditions do you 〜 ? .(38), (43) What conditions do you 〜 ? (43)

どのような試薬が必要になりますか？	What kinds of reagent do you need？(27) What kinds of reagent are required？(27)
どのようなステップを行いますか？	What steps are required 〜？(61) What actions do you need to take 〜？(61)
どのような方法がありますか？	What methods exist to 〜？(72) What methods can be used 〜？(73)
どのような方法を用いますか？	What method is used for 〜？(69) How do you 〜？(69)
どのような目的で	What's the purpose of 〜？(70) What's the point of 〜？(70)
どのようなものがありますか？	What kinds of 〜？(47), (75) What types of 〜？(47), (75) What sort of 〜？(47) **Is there** (a variety of plasmid preparation kits) **available** 〜？(47)
（抗体の標識法には）どのようなものがありますか？	What **methods** are there **to label the antibodies** used in Western blotting？(75) **How do you label the antibodies** in Western blotting？(76) **What means** are employed **to label the antibodies** used in Western blotting？(76)
どのようなものでしょうか？	What are the proper 〜？(25) What kind of 〜？(25)
どのような割り合いで	What's the ratio of 〜？(39)
どのように扱いますか？	How should you deal with 〜？(39)
（配列は）どのように決定しますか？	**How do you design** the primer for DNA sequencing？(59) **What do you need to think about** when you **design** the primer？(59)
どのようにしてDNA結合活性を測定しますか？	**How does the electro- mobility shift assay determine** DNA binding activity？(73) **How does the electro- mobility shift assay measure** DNA binding activity？(73)
どのようにして増やしますか？	What method do you use to amplify〜？(53)
どのようにして（〜を）集めればよいですか？	How do you collect 〜？(26) Where do you find 〜？(26) How do you search for 〜？(26)
（〜した後は）どのようにしますか？	**What's the next step after** centrifugation？(38) **What do you do after** you (do something)？(38) **What do you do once** you (have done something)？(38)
どのように対処しますか？	How do you handle 〜？(41), (46) What's the first thing you do 〜？(41) What do you do about 〜？(47)
どのように保存しますか？	**Under what conditions is** the media **stored**？(41) **How do you store** the media？(41) **What conditions do you store** the media under？(41)
ドラフトチャンバー	laminar hood (17)
トランスフォーム後，液体培地で増やし，DNAを精製します	You **perform transformation**, amplification in liquid media, and purification of DNA. (55) First **you transfer it**, then you amplify it in liquid media, and

	then you purify DNA. (55)
ナ 行	
流し（シンク）	sink (13), (105)
何を試してみますか？	what should you try to do? (78)
	what else can you do? (79)
（スタンダードな方法は） 何ですか？	What are the standard assays 〜 ? (71)
	What kind of standard protein assays 〜 ? (71)
肉眼で	with the naked eye (100)
	without the aid of a microscope (100)
濁る	turbid (91)
濃度はどれくらいですか？	What's the percentage of 〜 ?(50)
	What's the concentration of 〜 ?(50)
	At what concentrations should we add fetal bovine serum to the medium? (85)
	What percentage of fetal bovine serum should be added to the medium? (85)
延ばす所	spreading area (17)
ハ 行	
廃液処理	waste disposal (32)
廃液はどのように処理しますか？	How should you dispose of 〜 ?(33)
	How should you handle 〜 ?(33)
廃液用のプラスティックボトル	plastic bottle for waste liquid (105)
バイオハザード	biohazard (32)
培地の色の変化	changes in the color of the medium (86)
培地のpHの変化	change in pH levels in the medium (86)
培養液	liquid medium (32)
培養液入り瓶	bottle for culture media (16)
培養恒温器	incubator (16), (17)
（37℃のシェーカー内で） 培養します	Liquid media are incubated in a shaker at 37℃. (43)
	They're incubated in a shaker at 37℃. (43)
培養ディッシュ	culture dish (16), (93), (99)
培養ビン	culture bottle (84), (105)
培養フラスコ	culture flask (16), (90)
（96穴）培養プレート	96-well culture plate (99)
剥がれにくい	cells are not easily detached (94)
バキュームスイッチ	vaccum switch (14)
白衣	lab coat (17)
白衣掛け	coat rack (17)
バクテリア	bacteria (40)
パスツールピペット接続口	connector to Pasteur pipette (16)
白金耳	platinum loop (17)
バッファー	buffer (59)
（抽出してきたタンパク質の質に問題がなかったか） 判断できるね	You'll know whether the protein you extracted had a problem or not, right？(118)

バンド	band (63)
バンドを切り出してから	after cutting out a band (52)
	after you remove a band (52)
反応時間	reaction time (incubation) (46)
	time needed for a reaction (incubation) (46)
ビーカー	beaker (13), (69)
ビーカー設置板	beaker pad/beaker plate (13)
ヒートブロック	heating block (14)
必要な温度設定	temperature settings need to be made? (60)
	the required temperature settings? (60)
必要なものを挙げてください	Tell me what you need. (93)
	What do you need？(93)
ピペッター（ピペットマン）	pipette (pipetman) (12)
ピペッターとの接続部分	pipette connector (12)
ピペッティング	pipetting (19)
ピペットマン	pipetman (49), (78)
微量遠心機	micro-centrifuge (12)
ファージ	phage (40)
フィルターを通して濾過滅菌する	It should be filter sterilized. (84)
フィルターを通す	pass them through (97)
	filter them through (97)
フェノール	phenol (32)
フェノール層	phenol layer (38)
（抗原性を）賦活化する	activate its antigenicity (79)
	enhance its antigenicity (79)
不純物	contamination, impurities (70)
蓋	cover (16)
無難です	It is safer (88)
	It is wiser (89)
増やす	amplify, grow, multiply, culture (42)
（液体培地で）増やす	amplification in liquid media (54)
	culture in liquid media (54)
プラーク	plaque (40)
プライマー	primer (59)
フラスコ	flask (43)
（〜に）触れる	come into contact with 〜 (92)
プログラムフリーザー	programmed freezer (22)
ブロック	block (14)
ブロッティング装置	blotter (14)
プロテインA（プロテインG）アガロース	protein A (protein G) agarose (75)
プロテインA（プロテインG）セファロース	protein A (protein G) sepharose (75)
（−80℃の）フリーザー	-80℃ freezer (102)
分子式／示性式	structure (27)
分子量	molecular weight (27)
分子量マーカー以外全くバ	There are no bands other than those of the molecular weight

ンドを認めませんでした	markers. (114)
分注する	divide it up into separate (107)
	make aliquots (107)
粉末から作製した	that has been prepared from powder (84)
	made from powder (84)
ベクター (52)	vector (52)
(購入した少量の) ベクター	a small amount of commercially available vector (53)
	a small amount of commercially available vector (54)
	a small amount of vector purchased from a supplier (54)
	a small amount of vector procured from a supplier (54)
変性アガロースゲル	denaturing agarose gel (20)
放射性同位元素	radioisotope (30)
他にどんな方法がありますか？	What else can I do? (116)
他に予約は入ってる？	Has anyone reserved it? (115)
ボトルトップフィルター	bottle top filter (84)
ホモロジー検索を行う	carry out a homology search (67)
ポリアクリルアミドゲル	polyacrylamide gel (73)
ボルテックス	vortex (12)
ボルテックスをかける	vibrating on a vortex (19)
ホルムアルデヒド変性ゲル	agarose formaldehyde gel (20)
ポンプ	pump (46)
マ 行	
麻酔薬	anesthetic (30)
まずどのようなことをチェックしますか？	What should you check first (90)
	What's the first thing to look for (90)
(スターラーで) 混ぜる	mixing on a stirrer (19)
(〜の)ままで使用しますか？	Can you use PBS and trypsin solution kept at 4℃ as is? (93)
	If PBS and trypsin solution have been kept at 4℃, can you use them as they are? (94)
ミキサー	mixer (12)
水を入れるトレイ	water reservoir (17)
メスシリンダー	graduated cylinder (13)
目盛り	scale (12), (13)
目盛り調節ねじ	scale adjustment screw (12)
免疫沈降法	immunoprecipitation (75)
綿栓	cotton plug (43)
面にワセリンを塗る	apply Vaseline over (100)
メンブレン	membranes (21)
(〜で)用いる酵素は何ですか？	What kind of enzyme is used in foot print analysis? (74)
	What enzymes are used in foot print analysis? (74)
持ち手	grip (17)
(DNAの質が悪かったか，プライマーのデザインに)問題があったかもしれないね	The quality of DNA may have been bad, or there may have been some problems in the primer design. (115)
問題点は何ですか？	What problems might arise? (79)
	What points do you have to bear in mind? (79)

ヤ 行	
薬包紙	weighing paper (13)
融解する	thaw, defrost, unfreeze (87)
有機溶媒	organic solvent (30), (32)
遊離のDNAプローブ	free DNA probe (73)
指ではじく	tapping (19)
容量	volume (27)
予想サイズ	predicted size (56)
予想では2本出るはずのバンドが，1本しか出ない．	I thought there would be two bands, but there's only one. (113)
予約する	to reserve (28) to put your name down for (28) make a reservation for (28)

ラ 行	
(〜を用いる)利点は何ですか？	What is the advantage of using 〜 ? (48) What benefit is there to using 〜 ? (48)
粒子が大きくならないようにします	trying to avoid making large particles (57) taking care not to allow large particles to form (57)
利用できるコンピュータソフトがありますか？	Is there any computer software available? (59) Is there any computer software you can use? (60)
(ベンチの上に)両肘をついてサンプルを入れると入れやすくなります	You can load samples into each well more easily if you rest your elbows on the bench. (51) Placing your elbows on the bench top makes loading the samples easier. (51) Loading the samples is easier if you place your elbows on the bench top. (51) Placing one's elbows on the bench top facilitates easier gel loading. (51)
リンカー	linker (52)
リンカーの付加による正しい挿入	correct insertion by linker ligation (52)
凍結，融解を繰り返す	freeze-thaw cycles (69)
凍結細胞バイアル用コンテナ	cryocontainer (102)
凍結切片スライド	frozen section on a slide (78)
凍結切片の作製機器（クライオスタット）	apparatus to prepare frozen section (cryostat) (78)
冷蔵庫	refrigerator (22)
冷凍庫	freezer (22)
冷凍細胞バイアル	frozen cell vials (87)
ローター	rotor (12), (14)
ローテーター	rotator (75)
濾過滅菌水	filtered, sterilized water (18)

ワ 行	
（1：1の）割合	The ratio is 1:1. (39)
（1：3くらいの）割合で播く	seed at a ratio of about 1 to 3 (95) the ratio should be about 1 to 3 (95)

◆ **著者プロフィール**

村澤　聡（Satoshi Murasawa）

先端医療センター 再生医療研究部 主任研究員

1989年 関西医科大学医学部卒業，3年間の臨床従事のあと，同大学院博士課程入学．1996年 関西医科大学医学部博士課程修了（Ph.D）．1997～2000年 日本学術振興会特別研究員（PD）．2000年～2002年 米国タフツ大学セントエリザベス メディカルセンター留学（日本学術振興会海外特別研究員）．2002年4月より現職．
2001年11月 Young Investigator Award, Finalist of Louis N and Arnold M. Katz Research Prize, American Heart Association, USA. 2004年10月若手研究者奨励賞（Young Investigator Award）－優秀賞（基礎部門）受賞，第45回日本脈管学会総会．

大学院では，循環器内科の高血圧，動脈硬化部門に所属していたこともありアンギオテンシン受容体の分子生物学的機能解析の研究を行う．渡米前から，再生医療に興味を持つようになり，当時血管再生の遺伝子治療で世界的に有名だったJeffrey M. Isner教授のもとへ留学．同ラボにて血管内皮前駆細胞を発見した浅原先生（現 先端医療センター 再生医療研究部　部長）に出会い，血管内皮前駆細胞の機能解析と臨床応用についての研究を行う．帰国後，同細胞移植による虚血性疾患患者への治療応用のため，現在準備を進めている．

◆ **英文監修**：Geoff T. Rupp（Language Resources Ltd.）

バイオ実験　英語でトライ！
バイオ研究者のための基本英会話

2003年7月1日　第1刷発行
2008年5月30日　第3刷発行

著　者　村澤　聡
発行人　一戸　裕子
発行所　株式会社 羊　土　社
〒101-0052
東京都千代田区神田小川町2-5-1
TEL：03（5282）1211
FAX：03（5282）1212
E-mail：eigyo@yodosha.co.jp
URL：http://www.yodosha.co.jp/

装丁＆イラスト　ダイエイクリエイト 末永 弘二
印刷所　株式会社 平河工業社

©Satoshi Murasawa, 2003.
Printed in Japan
ISBN978-4-89706-368-3

本書の複写権・複製権・転載権・翻訳権・データベースへの取り込みおよび送信（送信可能化権を含む）・上映権・譲渡権は，（株）羊土社が保有します．

JCLS　＜（株）日本著作出版管理システム委託出版物＞ 本書の無断複写は著作権法上での例外を除き禁じられています．複写される場合は，そのつど事前に（株）日本著作出版管理システム（TEL 03-3817-5670, FAX 03-3815-8199）の許諾を得てください．

羊土社の大好評 英語学習関連書籍！

バイオサイエンス研究留学を成功させる
とっさに使える英会話
- 留学先のラボ
- 国際学会で そのまま使えるフレーズ集

★ジャンル別・シチュエーション別ですぐに見つかる対訳方式

オーディオCD付き

東原 和成／著　Jennifer Ito（東京大学大学院）／ナレーター
- 定価（本体 3,900円＋税）
- A5判　■ 192頁　■ ISBN978-4-89706-655-4

あのベストセラー「絶対話せる英会話」待望の続編．留学先ですぐに使えるフレーズや実験器具／研究用語を多数収録．さらに，会話・フレーズを収録したオーディオCDが付属．研究留学の強い味方！

医学・生物学研究者のための
絶対話せる英会話
研究室内の日常会話から国際学会発表まで

★バイオサイエンス用語の正しい発音からグラフの説明のしかたまでノウハウいっぱい

CD-ROM付き
MAC & WIN対応

東原 和成／著　Jennifer Ito（東京大学大学院）／ナレーター
- 定価（本体 3,950円＋税）
- A5判　■ 238頁　■ ISBN978-4-89706-629-5

研究室・セミナー・学会で困ったときのバイオサイエンス英会話集．日本人が間違えて発音している語句もリストアップされている．CD-ROMにはネイティブによる正確な発音や研究発表デモなどを収録！

ライフサイエンス
必須英和辞典

ライフサイエンス辞書プロジェクト／編著
- 定価（本体 3,800円＋税）
- B6変型判　■ 413頁　■ ISBN978-4-89706-484-0

PubMedの90％をカバー！ 生命科学の論文を読むならこの一冊でOK！手のひらサイズで使いやすい！！
すべての単語は和文索引からも引けて和英辞書としても使える！！

発行 **羊土社**

〒101-0052
東京都千代田区神田小川町2-5-1
TEL 03(5282)1211
E-mail：eigyo@yodosha.co.jp
FAX 03(5282)1212
URL：http://www.yodosha.co.jp/

ご注文は最寄りの書店，または小社営業部まで

論文執筆・学会発表などに役立つ英語関連書籍

ライフサイエンス 論文作成のための**英文法**

編集／河本　健
監修／ライフサイエンス辞書プロジェクト

約3,000万語の論文データベースを徹底分析！論文執筆でよく使われる文法が一目でわかる．「前置詞の使い分け」など，避けては通れない重要表現も多数収録．"初めて論文を書く学生も研究者も，これだけは押さえておきたい一冊！

■ 定価（本体 3,800円＋税）　■ B6判　■ 294頁　■ ISBN978-4-7581-0836-2

ライフサイエンス **英語表現使い分け辞典**

編集／河本　健，大武　博
監修／ライフサイエンス辞書プロジェクト

論文英語のフレーズや熟語を使いこなそう！ネイティブが執筆した約15万件の論文から得られた例文が満載で，「論文でよく使われる単語の組み合わせは？」といった，誰もが抱く論文執筆の悩みを解消する必携の一冊！

■ 定価（本体 6,500円＋税）　■ B6判　■ 1118頁　■ ISBN978-4-7581-0835-5

ライフサイエンス英語 **類語使い分け辞典**

編集／河本　健
監修／ライフサイエンス辞書プロジェクト

日本人が判断しにくい類語の使い分けを，約15万件の英語科学論文データ（全て米英国より発表分）に基づき分析．ネイティブの使う単語・表現が詰まっています．論文から引用した生の例文も満載で，必ず役立つ一冊！

■ 定価（本体 4,800円＋税）　■ B6判　■ 510頁　■ ISBN978-4-7581-0801-0

発行　**羊土社**

〒101-0052
東京都千代田区神田小川町2-5-1
TEL 03(5282)1211
E-mail: eigyo@yodosha.co.jp
FAX 03(5282)1212
URL: http://www.yodosha.co.jp/

ご注文は最寄りの書店，または小社営業部まで

論文執筆・学会発表などに役立つ英語関連書籍

困った状況も切り抜ける
医師・科学者の英会話

国際学会や海外ラボでの会話術と苦情,断り,抗議など厄介な対人関係に対処する表現法

著者／Ann M. Körner　訳・編／瀬野悍二

コミュニケーションがうまくいく,スマートな英会話を身につけよう! オーディオCD付き!

■ 定価（本体3,600円＋税）　■ B5変型判　■ 148頁　■ ISBN978-4-7581-0834-8

相手の心を動かす
英文手紙とe-mailの効果的な書き方

理系研究者のための好感をもたれる表現の解説と例文集

著者／Ann M. Körner　訳・編／瀬野悍二

相手に好印象を与え微妙なニュアンスが伝わる英文手紙の表現を解説! 豊富な例文収録のCD-ROM付き!

■ 定価（本体3,800円＋税）　■ B5変型判　■ 198頁　■ ISBN978-4-89706-489-5

日本人研究者が間違えやすい
英語科学論文の正しい書き方

アクセプトされるための論文の執筆から投稿・採択までの大切な実践ポイント

著者／Ann M. Körner　訳・編／瀬野悍二

20年間,7,000編もの科学論文の査読をしてきた一流研究者が授ける英語論文の書き方,免許皆伝!

■ 定価（本体2,600円＋税）　■ B5変型判　■ 150頁　■ ISBN978-4-89706-486-4

発行　羊土社
〒101-0052
東京都千代田区神田小川町2-5-1
TEL 03(5282)1211
E-mail: eigyo@yodosha.co.jp
FAX 03(5282)1212
URL: http://www.yodosha.co.jp

ご注文は最寄りの書店,または小社営業部まで